Uranium Enrichment and
Nuclear Weapon Proliferation

sipri

Stockholm International Peace Research Institute

SIPRI is an independent institute for research into problems of peace and conflict, especially those of arms control and disarmament. It was established in 1966 to commemorate Sweden's 150 years of unbroken peace.

The institute is financed by the Swedish Parliament. The staff, the Governing Board and the Scientific Council are international.

Governing Board

Dr Rolf Björnerstedt, Chairman (Sweden)
Dr Egon Bahr (FR Germany)
Mr Tim Greve (Norway)
Dr Max Jakobson (Finland)
Professor Karlheinz Lohs
   (German Democratic Republic)
The Director

Director

Mr Frank Blackaby (United Kingdom)

sipri

Stockholm International Peace Research Institute
Bergshamra, S-171 73 Solna, Sweden
Cable: Peaceresearch, Stockholm
Telephone: 08-55 97 00

Uranium Enrichment and Nuclear Weapon Proliferation

Allan S. Krass
Peter Boskma
Boelie Elzen
Wim A. Smit

sipri

Stockholm International Peace Research Institute

International Publications Service
Taylor & Francis Inc.
New York
1983

First published 1983 by International Publications Service
Taylor & Francis Inc., 114 East 32nd Street
New York, NY 10016

Typeset in the United Kingdom by C.P. Group,
Croydon, Surrey
Printed in the United Kingdom by
Taylor & Francis (Printers) Ltd, Rankine Road,
Basingstoke, Hampshire RG24 0PR

Library of Congress Cataloging in Publication Data

Main entry under title:

Uranium enrichment and nuclear weapon proliferation.

   Stockholm International Peace Research Institute. Includes
bibliographical references and index.
   1. Uranium enrichment. 2. Atomic weapons. I. Krass, Allan S.
II. Stockholm International Peace Research Institute.
TK9360.U735   1983   327.1'74   83-8486
ISBN 0-8002-3079-5

Preface

In the years following World War II, when gaseous diffusion was the only practical means of enriching uranium, the potential contribution of uranium enrichment to weapon proliferation seemed small. It was assumed that gaseous diffusion plants were so technically complex and costly that only the most wealthy, industrialized countries could afford them. The potential for the diversion of plutonium to weapon use seemed far more dangerous and received far more attention from people concerned about weapon proliferation.

However, during the past 20 years there has been a steadily accelerating technological advance across the broad spectrum of enrichment technology. These developments have made the enrichment route to proliferation more accessible to smaller, poorer and less technically advanced countries. This has in no way reduced the danger associated with plutonium, but has instead added a new set of dangers and problems for those who would attempt to stop and reverse the spread of nuclear weapons.

This book presents a comprehensive review of the state of the art of enrichment technology and attempts to evaluate the impact of this technology on the proliferation problem. It places the technical development into the context of the economic and institutional environment within which the enrichment industry has evolved to its present state, and it suggests some measures which might be taken to reduce the proliferation dangers inherent in this industry.

Part of this book was prepared in Sweden and part in the Netherlands. The total work was co-ordinated at SIPRI in Stockholm by Allan Krass, who is Professor of Physics and Science Policy at Hampshire College (Amherst, Massachusetts, USA). Dr Peter Boskma holds the chair of Philosophy of Science and Technology and Dr Wim A. Smit is Director of 'de Boerderij' Centre for Studies on Problems of Science and Society, both at Twente University of Technology (Enschede, the Netherlands). Mr Boelie Elzen is a research fellow at 'de Boerderij' Centre and holds a university degree in electrotechnical engineering. Special gratitude is

v

extended to Mr Gerard W.M. Tiemessen, who is also a research fellow at 'de Boerderij' Centre, and who has made substantial contributions to parts of this book; he holds a university degree in physical engineering.

Acknowledgements

Acknowledgement is given to Barbara Adams and Connie Wall for editing the manuscript, and to Alberto Izquierdo for drawing the figures.

Special acknowledgement is also given to the library staff at Studsvik Energiteknik, Nyköping, Sweden, particularly Lars Edvardson and Maud Sundblad, for their very significant help and co-operation.

SIPRI   *Frank Blackaby*
1982   Director

Contents

PART TWO

List of figures and tables

PART THREE

Chapter 7. A history of non-proliferation efforts

Chapter 8. The world enrichment picture

Appendix 8A. Nuclear power growth 1980–1990

Appendix 8B. Demand for and supply of enrichment services

Introduction

This book is concerned with the connections between uranium enrichment technology and the horizontal proliferation of nuclear weapons. It is intended to provide journalists, policy analysts, diplomats, students, and political and bureaucratic decision-makers with detailed and complete descriptions and analyses of a large number of important developments, which during the past decade have greatly increased the role which enrichment is playing and will continue to play in the proliferation problem.

Uranium enrichment is one of two methods which have been developed to produce nuclear explosive material. The other is the production of plutonium. Although plutonium production and its attendant proliferation problems are not dealt with in this book, this should in no way be interpreted as a denigration of their importance. Any successful non-proliferation strategy will have to deal with the problems presented by both paths, but a complete study of both is beyond the capabilities of the present authors.

For many years uranium enrichment was carried out by means of one technique, gaseous diffusion, and predominantly in one country, the USA. China, France, the UK and the USSR had by far the largest capacity and for many years held the dominant position in the world enrichment market. However, starting in the mid-1970s this dominant position began to decline slowly as other countries entered the market and as new techniques became more competitive with gaseous diffusion.

It was the anticipation in the 1960s of a large and growing world market for enrichment services which stimulated the research and development (R&D) programmes which led to the new techniques. Most people believed at that time that nuclear energy would spread throughout the world and that the dominant power reactor design would be based on the light-water moderator and coolant, a design which requires enriched uranium for its fuel. The combination of prospects for a growing enrichment market and a growing desire for energy independence led many countries to invest significant resources in an attempt to acquire an indigenous enrichment capability. It is not unreasonable to assume that some portion of this effort was also motivated by a desire to keep an option to produce nuclear weapons.

During this same period concern over weapon proliferation relaxed somewhat. The signing of the Non-Proliferation Treaty (NPT) and the establishment of an international safeguards system led many to believe that this problem could be brought under control. Of course, it was recognized that many of the features which made such new enrichment techniques as the gas centrifuge and the jet nozzle processes commercially attractive also made them vulnerable to acquisition or misuse by states wanting to acquire nuclear weapons, but concern over the problem was muted by optimism that adequate safeguards arrangements could be worked out.

Unfortunately, as the 1980s began, these optimistic assumptions appeared to have been premature. A deep recession in the world nuclear power industry has left many countries with an over-capacity of enrichment capability and large capital investments in research, development, and plant construction which now threaten to be unrecoverable. A number of processes have been developed which clearly make proliferation easier, and these, coupled with the continuing desires of some countries for self-sufficient nuclear fuel cycles and the desires of others to export technology, have produced a highly unstable situation with respect to proliferation.

Given these pressures for the spread of new enrichment techniques, it is then doubly disturbing to discover that the various institutional and technical mechanisms for safeguarding enrichment facilities and for controlling exports of sensitive components and know-how are seriously inadequate. Very little experience exists in applying the usual safeguards to enrichment facilities, and both technical and institutional obstacles must be overcome before genuine confidence can exist that enrichment facilities and their product can be monitored with confidence.

All of this leads to the conclusion that the proliferation implications of uranium enrichment technology must be understood much better and taken much more seriously than they have been in the past. The purpose of this book is to gather together in one place the material necessary for such an in-depth analysis.

The book is divided into three parts. In Part One the connections between enrichment and nuclear weapon proliferation are explored and analysed. The argument is presented without technical detail or extensive data on enrichment demands and capabilities. These are supplied in Parts Two and Three.

The argument of Part One proceeds from a brief description of the basic physical principles of uranium enrichment (chapter 1) to a comparative analysis of the proliferation potentials of various processes and the capabilities and intentions of the many nations that have demonstrated interest in enrichment (chapter 2). The problems of control raised by enrichment developments are explored in chapter 3 both from technological and institutional points of view. Then in chapter 4 we outline the conclusions of our analysis and make a number of policy recommendations whose aim is to retard the further proliferation of nuclear weapons, at least until the incentives to acquire them can be reduced or eliminated.

Part Two is intended for a more technically oriented audience. It describes the scientific and technological aspects of enrichment in sufficient detail to allow independent analysis and evaluation of both present and future technological developments. The significant parameters of the various processes are summarized, and a framework is provided for understanding the proliferation implications of these parameters. Part Two presents all of the technical data necessary to support the arguments of Part One.

Part Three begins with a brief history of efforts to control nuclear weapon proliferation with an emphasis on the role played by enrichment (chapter 7). It is shown that the enrichment industry has played a dual role in both facilitating proliferation and providing mechanisms by which proliferation might be discouraged or retarded. This is followed by a detailed picture of the world enrichment industry and market — past, present and future. A country-by-country summary of enrichment capabilities, intentions and non-proliferation attitudes is presented and summarized in a number of tables and appendices. As with Part Two, these data provide the basis for many of the assessments and conclusions made in Part One.

NOTES ON CONVENTIONS

The following general conventions are used in the tables:
.. Information not available
( ) Uncertain data or SIPRI estimate
– Nil or not applicable

'Billion' in all cases is used to mean thousand million.

All dollar prices are US dollars.

Metric units generally apply:
1 tonne (metric ton) = 1 000 kilograms = 2 205 pounds = 1.1 short tons
1 megawatt (MW) = 10^6 watts
1 gigawatt (GW) = 10^9 watts

Part
One

Chapter 1. Fundamentals of uranium enrichment

I. Isotopes

The atoms of all chemical elements consist of a very small nucleus surrounded by a cloud of electrons. The nucleus accounts for all but a tiny fraction of the mass of the atom, but is confined to a volume whose diameter is only one ten-thousandth of the diameter of the atom itself. Since all the reactions between atoms, that is, all of chemistry, take place at the outer fringes of the electron cloud, it is not difficult to understand why the structure of the nucleus, aside from its total electric charge, has only very minor effects on chemical processes.

The structure of the nucleus is determined by the numbers of its two constituents, protons and neutrons. Figure 1.1 shows a graph of the stable, naturally occurring elements in which the number of neutrons, N, is plotted against the number of protons, or atomic number, Z [1]. It is the latter value which determines how the element behaves chemically, while the former affects only the mass and detailed structure of the nucleus. A chemical element is determined by its value of Z, and an element can have several isotopes characterized by different values of N.

Figure 1.1 shows that not all combinations of N and Z are possible, and that the stable isotopes are confined to a relatively narrow band of stability. For low numbers of protons the number of neutrons tends to be close to Z, but as the number of protons increases, the number of neutrons needed to ensure stability increases more rapidly. For example, in a stable isotope of lead there are 82 protons and 126 neutrons, giving a total of 208 particles. This total number is often designated A and is called the atomic weight. Since protons and neutrons have very nearly equal masses, this value of A is a quite accurate measure of the total mass of the nucleus, and therefore of the entire atom.

The physical explanation for the presence of more neutrons than protons in heavy nuclei lies in the role neutrons play in holding the nucleus

1

Figure 1.1.The stable isotopes

The stable isotopes are indicated by dots whose co-ordinates are the neutron number N and the proton number Z. Note the increasing ratio of neutrons to protons for heavier nuclei. The heaviest stable isotope is ^{209}Bi, and the isotopes of thorium and uranium shown are present in nature only because of their very long lifetimes.

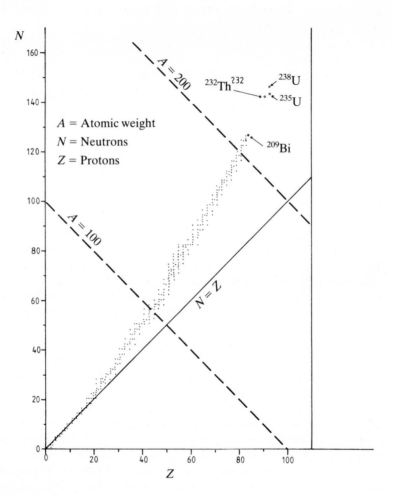

together. Stability in the nucleus is actually the result of a delicate balancing of several forces, the two most important of which are the attractive nuclear force and the repulsive electrical force. Extra neutrons are needed to provide sufficient attractive forces to overcome the mutual repulsion of the protons. By the time the number of protons reaches 80 this balancing becomes quite difficult and, as figure 1.1 shows, there are no stable nuclei beyond $Z = 83$, the single stable isotope of bismuth.

If all elements above bismuth are unstable, then how can the existence of natural uranium (with 92 protons) be accounted for? The explanation is in the fact that the instability of nuclei can manifest itself over a wide range of lifetimes. There happen to be three particular combinations of protons and neutrons which are so nearly stable that their half-lives (the time in which half of a given sample will decay) are of the order of hundreds of millions or billions of years, comparable to the age of the Earth. These three are: one isotope of thorium, ^{232}Th (with a half-life of 14 billion years); and two isotopes of uranium, ^{235}U (with a half-life of 710 million years) and ^{238}U (with a half-life of 4.5 billion years). These species are also indicated in figure 1.1.

The generally accepted theory of the formation of the Earth is that it condensed from a cloud of material, possibly the remnants of an exploded star. At the time of the Earth's formation the relative amounts of all the elements and their isotopes were determined by whatever nuclear reactions had accompanied the formation of the cloud; but since the formation of the Earth some five billion years ago there has been no significant creation of new heavy elements, and those that solidified with the Earth have been decaying away. After five billion years only the three above-mentioned species remain in any significant quantities. The other heavy elements which are found (e.g., the radium discovered by Marie Curie) are present as a result of the decay of these three.

If the half-lives of the two uranium isotopes[1] are compared with the life of the Earth, an interesting result emerges. First, notice that the age of the Earth is slightly longer than one half-life of ^{238}U. This means that in the early years of the Earth's life there was roughly twice as much ^{238}U present as there is now. But ^{235}U has endured over seven half-lives in the same period, which implies that in the early days there was about 2^7 or 128 times as much as there is now. At present natural uranium consists of 99.3 per cent ^{238}U and only 0.7 per cent ^{235}U, but if these values are extrapolated backwards in time, it can be shown that primordial natural uranium must have contained almost 30 per cent ^{235}U.

From this it is clear that if the Earth had been inhabited soon after its formation, nuclear energy would have been quickly and easily discovered. Actually, long before human life appeared on the planet, a natural nuclear reactor went critical in a rich uranium seam under what is now Gabon in West Africa [2]. This event took place about 1.7 billion years ago when natural uranium was slightly more than 3 per cent enriched in ^{235}U, a value practically identical to that employed today in most light water nuclear reactors (LWRs). This book deals with the elaborate and costly methods which have had to be devised to make up for the arrival of human beings two billion years too late to take advantage of this situation.

An important mechanism of radioactive decay in heavy nuclei is nuclear fission. In this process the electrical repulsive forces win out over

---

[1]   A third isotope, ^{234}U, is found in very small quantities in natural uranium. This isotope is too short-lived to have survived since the Earth's formation. It may be a minor product of the decay of ^{238}U.

the nuclear attractive forces, and the nucleus breaks apart into two or more pieces, usually accompanied by the release of a few neutrons. This fission can take place either spontaneously, like other radioactive processes, or it can be induced by the absorption of energy or another particle. There are several nuclei which undergo fission after absorbing a neutron, but generally a rather energetic neutron is required to induce fission.

Of the three naturally occurring, long-lived nuclear species mentioned above, only one, ^{235}U, undergoes nuclear fission when bombarded with slow neutrons. This property is important for the maintenance and control of a sustained, as opposed to an explosive, nuclear chain reaction (i.e., a reaction which is capable of sustaining itself once it has been started). The other two species, ^{232}Th and ^{238}U, can absorb neutrons and be converted (bred) into other forms (^{233}U and ^{239}Pu) which also fission in the presence of slow neutrons, but this would probably never have been more than a scientific curiosity if ^{235}U had not been available in natural form. Only the chain reaction in ^{235}U could have provided at feasible costs the enormous numbers of neutrons needed to breed sufficient quantities of the other isotopes to begin a nuclear fuel cycle or to produce nuclear weapons. Depending on how one assesses the impact of nuclear energy on human history, one can see this remarkably anomalous phenomenon as a stroke of either extremely good or extremely bad luck.

It is possible to maintain a chain reaction in natural uranium. This was in fact what was done by all the countries which contributed to the early development of nuclear energy – Canada, France, the UK, the USA and the USSR. The very small concentration of ^{235}U in natural uranium enforces very stringent requirements for neutron concentration in such a reactor. Neutrons emitted in a fission reaction must be slowed down by many collisions with nuclei in the surrounding medium, called the moderator. If too many are absorbed during the slowing down process, not enough will be left to sustain the chain reaction. Only very pure carbon (graphite) or heavy water $(D_2O)^2$ have been found sufficiently resistant to neutron absorption, and producible in sufficient quantities at acceptable cost, to serve as moderators in natural uranium reactors.

It was realized very soon after the discovery of nuclear fission in 1939 that it might be possible to make a powerful nuclear explosive by extracting and concentrating the ^{235}U from natural uranium. It also became clear that if the uranium could be enriched in this isotope it would be possible to substitute ordinary water for the expensive graphite or heavy water in a reactor. These motivations, particularly the first, provided the early impetus for devising means for producing enriched uranium. Later on it was learned that very pure ^{235}U was also extremely useful, if not essential, for the design of thermonuclear explosives. Given these objectives, and the world political situation which emerged from World War II, it is not

---

2 "D" is the symbol for deuterium, an isotope of hydrogen containing one proton and one neutron.

surprising that a great deal of scientific and technological creativity was devoted to the problem of enriching uranium. This book is concerned with some of the results and implications of this creativity.

II. Uses for enriched uranium

As has already been mentioned, even natural uranium can be useful in certain types of nuclear reactors, but, as the proportion of ^{235}U is increased, the ability to use uranium as an energy source becomes more flexible and technically less complicated. Modern light water reactors use uranium enriched to about 3 per cent, although this can vary anywhere between 2.5 and 3.5 per cent depending on the type and operating schedule of the reactor.

As the percentage of ^{235}U increases, the size of reactor for a given power level can decrease, that is, the power density in the core of the reactor increases. Reactors used for ship propulsion use enrichments of at least 10 per cent in order to keep the size of the power plant relatively small. Nuclear submarines, satellites, and many small research reactors use fuel enriched to 90 per cent or more [3a]. Because of the high power densities achievable with such enrichments, research reactors generally require only a few kilograms of fuel to achieve criticality (i.e., a self-sustained nuclear reaction).

Below 10 per cent enrichment metallic uranium cannot be made to explode, since the mass which must be assembled for the explosion, the so-called critical mass, is essentially infinite [4a]. However, the critical mass drops rapidly above 10 per cent enrichment, especially if the uranium is surrounded by a good neutron reflector. This material serves to return neutrons which would otherwise be lost back into the uranium to create more fission reactions. It can also serve to hold the critical mass together for a longer time in order to increase the power of the explosion [5].

Table 1.1. Critical masses of uranium for varying degrees of enrichment[a]

Enrichment (per cent)	Critical mass of uranium-235 (kg)
10	100
20	50
50	25
100	15

[a] The critical mass values, which assume a good neutron reflector, are taken from reference [4b].

5

By the time enrichment reaches 20 per cent, uranium can be considered a highly sensitive material. This enrichment has in fact been chosen by the International Atomic Energy Agency (IAEA) as the line above which enriched uranium is treated in the same category as plutonium, the other nuclear explosive material [6a]. Table 1.1 shows that the critical mass required for a nuclear weapon drops rapidly above enrichments of 20 per cent, and that even enrichments as low as 50 per cent must be viewed as potentially dangerous. However, as the rest of this chapter and the next will show, the production of even relatively small amounts of highly enriched uranium is no simple task.

III. Isotope separation

Uranium enrichment is a special case of the much more general problem of separation of isotopes. There are many applications, for example, in biological research and medicine, for which it is necessary or desirable to obtain purified samples of a particular isotope; but, generally the amounts required are extremely small: a gram of purified isotope is considered a large amount in a biological laboratory. Only for uranium and hydrogen have people found it necessary to perform isotope separation on quantities of material measured in millions of kilograms.

Since the number of neutrons in a nucleus has only minute effects on the chemistry of an atom, the separation of isotopes cannot be accomplished by the use of chemical techniques of the kind normally used to purify substances. The separation must somehow take advantage of those properties of the nucleus which are affected by the number of neutrons: its mass, size and shape, magnetic moment or angular momentum. By far the most important of these is the difference in mass, which is utilized in all of the proven enrichment processes and all but one of the advanced methods currently under development.

There are two broad classes of effects which are sensitive to the difference in mass of uranium isotopes. The first relies on the differences in the average speed of atoms or molecules with different masses which are mixed together in thermal equilibrium. The second utilizes the different inertias, that is, resistance to acceleration, of different masses when they are subjected to the same force. Lighter atoms or molecules move faster on the average than heavier ones in mixtures at a given temperature, and lighter molecules respond more readily (i.e., accelerate faster) than heavier ones when a force is applied to them.

In the first class can be placed all of the currently demonstrated separation techniques: gaseous diffusion, centrifugation and aerodynamic techniques such as the jet nozzle process and the vortex tube process. All of these processes operate on uranium in the form of uranium hexafluoride gas, UF_6.

At any temperature the molecules of a gas will be in constant motion, colliding with each other millions of times per second and exchanging energy each time they collide. All of these collisions ensure that the energy of motion of the molecules (their kinetic energy) is equally shared among them on the average. This equal sharing of average kinetic energy is one of the fundamental principles of the kinetic theory of gases, or more generally, the branch of physics called statistical mechanics which deals with systems containing very large numbers of particles.

If two particles have the same kinetic energy, the one with the smaller mass will have the larger velocity. In the gaseous diffusion process this difference is exploited by allowing the gas to diffuse through a solid barrier permeated by many small holes or pores. The faster-moving molecules pass through the holes more frequently, and the mixture which emerges on the other side of the barrier is therefore somewhat richer in the light species than the original sample.

The centrifuge and aerodynamic methods also use diffusion, but add strong accelerations to magnify the effect. In all of these methods the gas is accelerated into rapid rotation creating a centrifugal force field in the gas which accelerates particles towards the periphery of the circle.

This force has a property similar to gravity in that it accelerates particles at a rate which is independent of their masses. For example, when a car turns a sharp corner, everything and everybody in the car is thrown to the outside of the turn at the same rate. It follows that centrifugal force by itself cannot be used to separate isotopes. All it can do is create a pressure variation in the gas, against which particles of different masses will diffuse at different rates. It is this differential diffusion against a pressure gradient which underlies the centrifuge, nozzle and vortex tube techniques.

There are forces which do accelerate particles of different masses at different rates. In this class are electromagnetic forces, interatomic binding forces and the forces between colliding atoms or molecules. Actually all of these are electromagnetic in origin, but it is more convenient here to treat them separately.

If an atom or molecule can be ionized, that is, given an electric charge, it can then be accelerated by either electric or magnetic fields or both. The earliest successful uranium isotope separating device utilized strong magnetic fields to deflect beams of ^{235}U and ^{238}U ions by slightly different amounts. This device, called the calutron, produced the highly enriched ^{235}U which destroyed Hiroshima in 1945. Another technique which was suggested in the 1940s, but which has only recently begun to show some promise of success, is the use of oscillating electromagnetic fields to accelerate ionized atoms in circles, using a phenomenon called ion cyclotron resonance.

Interatomic forces hold molecules together, and a useful model of a chemical bond is a small spring connecting two masses. This spring can be stretched and compressed, so that the masses can oscillate with certain characteristic frequencies. These frequencies tend to lie in the infra-red part of the electromagnetic spectrum, and the vibrations can be excited either by collisions with other molecules or by absorption of infra-red radiation at well-defined frequencies. The vibration frequency depends on the stiffness of the spring (i.e., the nature of the chemical bond) and the values of the masses; but since the chemical bonds formed by two isotopes are virtually identical, the masses alone determine the vibration frequency shifts between molecules containing different isotopes.

The principle of molecular laser isotope separation (MLIS) is to tune an infra-red laser to a precise vibration frequency of $^{235}UF_6$. Then, only these molecules are excited into vibration, and once they are, they can be separated in any of a number of ways by ionization, dissociation, or chemical reactions. Recent proposals also suggest that this method of selective infra-red excitation can be used as a way of giving the molecules more thermal energy and utilizing this to enhance various diffusion effects. The difference in vibration frequencies also leads in a rather subtle way to a tendency for light isotopes to concentrate in certain chemical compounds in preference to others. To exploit this effect the two chemicals must be brought into contact with each other and encouraged, usually with catalysts, to exchange uranium atoms. When this exchange reaches

Table 1.2 Uranium enrichment techniques, according to physical principles and mechanisms used

Property of isotope	Physical principle	Physical mechanism	Separation process
Nuclear mass	Newton's law of acceleration	Acceleration by electro-magnetic forces	Calutron Plasma centrifuge Ion cyclotron resonance
	Quantum theory of molecular bonds	Shift in chemical equilibrium	Chemical exchange
		Selective absorption of infra-red light	Molecular laser isotope separation (MLIS)
	Equipartition of energy	Flow through porous barrier	Gaseous diffusion
		Diffusion against pressure gradient	Centrifuge Jet nozzle Vortex tube
Nuclear size, shape, spin, magnetism	Quantum theory of atomic structure	Selective absorption of visible or ultraviolet light	Atomic vapour laser isotope separation (AVLIS)

equilibrium a slight, but significant, concentration of isotopes is sometimes achievable. This is the basis of a number of chemical-exchange techniques which are currently under development.

There is one isotope separation process which does not depend very much on the nuclear mass difference. This is atomic vapour laser isotope separation (AVLIS) which utilizes very small shifts in the frequencies at which uranium atoms absorb light. These shifts are caused only partially by the different masses of the two nuclei; more important are the small changes the extra neutrons make in the magnetic properties of the nucleus as well as in its size and shape. These changes are very small, and it was only the advent of the laser which made possible the selective use of light to distinguish such small frequency shifts. The process uses laser light to excite and ionize only ^{235}U atoms in a chamber containing metallic uranium vapour. Then the ^{235}U ions are collected by electromagnetic fields.

Table 1.2 summarizes this brief introductory discussion by listing the enrichment techniques mentioned above along with the basic principles and mechanisms on which they are based. From this list it is clear that there is no shortage of clever ideas for the separation of uranium isotopes. The basic physical principles underlying most of the above-mentioned methods are really quite simple, and all have been well understood for many years. But the transition from an elegant physical principle to an economically viable industrial technique is usually a long and arduous one. The reasons for this are made clear in Part Two where the practical engineering aspects of these ideas are explored.

IV. Basic principles of enrichment

The basic component of any enrichment facility, no matter which process is used, is the enrichment element (see figure 5.1, p. 95). An element is a device which separates an incoming feed stream into two outgoing streams: a product stream in which the process material is enriched to some degree in the desired isotope and a tails stream (sometimes inappropriately called waste) which is somewhat depleted in this isotope.

The degree of separation which can be achieved in a given element is measured by a parameter called the separation factor. This varies widely from one technique to another, being smallest for the chemical-exchange processes and gaseous diffusion and largest for the modern resonance techniques based on lasers or plasmas. When the separation factor is very small only a small enrichment can be achieved in a single element, and the process material must be passed through many elements in order to achieve useful enrichments. For example, several thousand separate enrichment elements must be employed in a chemical-exchange facility to enrich uranium from its natural composition of 0.71 per cent ^{235}U to 3 per cent ^{235}U. On the other hand it is possible that an atomic vapour laser enrichment element will be able to obtain the same separation in a single element.

Another important property of any enrichment element is the rate at which it can process feed material. Some elements, such as large gaseous diffusion units, can process tons of material per minute, while others, such as those which use atomic vapour might process only a few grams per minute or even less. Both the separation factor and the flow rate are needed to specify the separative power of an element, which is usually rated in separative work units per year (SWU/yr). For example, typical ratings for modern, high-speed gas centrifuges are from 5 to 100 SWU/yr depending on their design and operating conditions.

An enrichment facility designed to produce fuel for several nuclear reactors must have a total capacity of hundreds of thousands of SWU per year. Such a plant must therefore consist of many elements arranged in a cascade. Such an arrangement is illustrated in figure 5.4, p. 103 which shows a series of stages each consisting of a number of elements. By adding elements to a given stage the flow capacity of the stage can be increased, and by cascading many stages in series any desired enrichment can be achieved in the final product.

In general, it can be said that the smaller the separation factor of a stage the larger the flow rate necessary to achieve a given separative capacity. This can be achieved either by inserting many individual elements in a stage (as is done with centrifuges) or by building very large elements which constitute a stage in themselves (as in gaseous diffusion). The more elements that are required, or the larger their sizes, the larger the size of the facility. Also, the larger and more numerous the elements become, the larger is the process inventory in the plant. Since the inventory must consist of uranium it is an important economic factor in the design of enrichment plants. Large plant sizes and process inventories also make it difficult to hide such facilities or to operate them in ways for which they were not designed.

For most enrichment techniques there is a time delay between when a plant is completed and when useful product can be extracted. The cascade must first be filled with unenriched feed material and then run for a while to allow the concentration of desired isotope to build up at the product end. A standard measure of the time required for this build-up is the equilibrium time of the cascade. This time depends on the final enrichment desired, on the inventory, and on the time required for a sample of material to pass through a typical element. It can range from virtually instantaneous for the laser and plasma processes, to weeks or months for plants with large inventories and small separation factors. An extreme example is the 30–40 year equilibrium time required for obtaining 90 per cent enriched product in one type of chemical-exchange facility.

An enrichment cascade is normally designed to produce product with some desired degree of enrichment from feed with some specified initial isotopic composition. This is usually natural uranium, but in principle a cascade can be fed with material of higher enrichment, and the product of such a cascade will be correspondingly more highly enriched. This leads to the possibility of batch recycling to produce highly enriched material. Batch recycling is generally a wasteful way of achieving high enrichments, but may be an option if obtaining relatively small amounts of highly enriched material is of high priority. The suitability of any technique for batch recycling is obviously related to its equilibrium time and inventory.

There are two more features of a cascade which should be introduced here in preparation for the comparative analysis of the next chapter. The first is reflux and the second is criticality.

Reflux is the process by which part or all of the output of an element is recycled back to the input to achieve higher separation factors. In effect this amounts to running the material through a given element more than once, and the ease with which this can be done depends on the nature of the process material and the enrichment mechanism. If, as in chemical-exchange techniques, the chemical nature of the process material is changed by the enrichment process, then the reflux must consist of a chemical reconversion. This adds considerably to the cost and technical difficulty of chemical enrichment techniques.

Criticality problems arise in any process where enough fissionable material might collect to produce a self-sustaining nuclear reaction (i.e., a critical mass). This is not a serious problem where the process inventory is in gaseous form, but when liquids (especially aqueous or organic solutions) are used, special precautions must be taken to make sure that no critical mass can ever accidently be assembled.

This completes a brief introduction to the basic principles and terminology of uranium enrichment. With this introduction the non-technical reader should be able to follow the arguments of the next chapter in which the proliferation dangers of the various processes are compared. Readers desiring a more detailed and quantitative discussion of these concepts are referred to Part Two.

Chapter 2. Enrichment and proliferation

I. Introduction

There has been remarkable progress in both the variety and efficiency of enrichment methods since the first primitive calutrons began operation in 1944. Now there exist two enrichment methods in commercial use and four more which have demonstrated their technical feasibility and which are either in extensive operation or show substantial promise of commercial feasibility. There are three more techniques which are under active research and development and for which vigorous efforts are being made to demonstrate their feasibility and competitiveness with the other methods. Finally, there is a generous pot-pourri of less-developed inventions, ideas and suggestions, any of which might quickly emerge as an important new process. This creative ferment has been stimulated partly by the scientific and commercial value of purified isotopes in general, but mainly by the particular value of enriched uranium for both civilian and military purposes.

The purpose of this chapter is to examine what effects all of this creativity has had and is likely to have on the possibility of proliferation of nuclear weapons. In order to do this properly it is necessary to consider both the technical aspects of the various enrichment methods and the historical, economic and political context into which they are being introduced. Therefore, this chapter carries out a technical analysis of the major enrichment techniques with an emphasis on their susceptibility to application to the production of highly enriched uranium which is suitable for nuclear explosives. This susceptibility is described in terms of technological thresholds to proliferation, where a low threshold implies a high susceptibility. Given the quantitative information available on the various techniques, these thresholds can be determined with reasonable accuracy, and the techniques are ranked according to their relative proliferation sensitivity.

This is followed by an analogous attempt to define situational thresholds, which apply to the various nations who either have, or are known to be interested in obtaining, enrichment capabilities. A situational threshold is a measure of the national and international political obstacles facing a nation thinking about acquiring an enrichment capability and using it to produce nuclear weapons. The definition of this threshold is made more specific in section IV of this chapter.

Situational thresholds are intended to be analogous to technological thresholds in that both are assumed to be more or less objectively determinable. By combining the two it is possible to obtain a measure of the degree to which the acquisition of certain technology by a particular state would represent a manifest threat to the non-proliferation regime. This can be done independently of any judgement regarding the motives of the state in question.

It is obvious that motives are also extremely important in estimating the risks of further proliferation of nuclear weapons. However, it is very difficult to determine motives to any reasonable degree of validity. There are those who would go so far as to say that it is wrong in principle to attribute purposeful motives to states. In this view the actions of states are really only the result of internal political or administrative processes in which parochial or bureaucratic interests are more important than any overall national purpose [7].

As interesting and useful as this approach may be in many circumstances, this study has chosen to assume that the result of such processes can appear as states having motives, and that these can be determined with some reliability by observing the consistency between the actions of a state and its professed national objectives. For example, the vast majority of non-nuclear weapon states assert strongly that they have no interest in acquiring nuclear weapons. Yet many such states have acted in ways which are not consistent with such assurances. Inconsistencies can be observed in a number of areas, including the relationship between research and development priorities and the general technological level of the state, the relationship between enrichment activities, energy independence policy or indigenous resource development, and in the state's attitudes towards the Non-Proliferation Treaty and international safeguards. None of these measures are infallible, but as will be demonstrated at the end of the chapter, when they are taken together they provide a quite plausible picture of the proliferation implications of recent and potential developments in uranium enrichment.

II. The origins of uranium enrichment

Uranium enrichment has a history very different from that of most industrial processes. It was developed in an atmosphere of intense urgency

13

and with virtually none of the normal constraints on cost, efficiency and profitability. In the USA during World War II there was only one goal: to produce enough nuclear explosive material to make a weapon or weapons which could be used in the war. This was achieved, but its achievement resulted in a highly distorted subsequent development of the enrichment industry.

The process which actually succeeded in reaching the objective was the calutron. It produced only enough highly enriched uranium to make one bomb, the bomb that destroyed Hiroshima. Only a few months later the calutrons were shut down and never again used to separate uranium in large quantities [8a].

The calutrons had been provided with slightly enriched feed material from two other processes. One was a thermal diffusion method developed at the US Naval Research Laboratories. In this process a temperature difference is created across an annular region between two long, vertical cylindrical pipes. The inner pipe carries high temperature steam, while the outer is cooled by circulating water. Uranium hexafluoride is kept in liquid form under pressure in the region between the pipes, and in a very slow process the lighter isotope tends to move towards the hot inner wall and then rise to the top of the tube. The heavier isotope diffuses outward and is collected at the bottom [8b]. This method can produce only very small enrichments, even in tubes as high as 48 feet (14.6 m). Nevertheless, a plant containing more than 2 000 such tubes was constructed at Oak Ridge, Tennessee in less than eight months in order to provide feed enriched to 0.89 per cent ^{235}U for the calutrons [8c]. Such a lavish expenditure of money for such a marginal achievement could only happen in an atmosphere of intense crisis. Less than a month after the bombing of Hiroshima the thermal diffusion plant was shut down [8d].

The other process feeding the early calutrons was gaseous diffusion, which survived World War II and went on to dominate the industry totally for the next 30 years. The scale and expense of the effort expended to develop this process makes the previous two pale by comparison. Consider the following description of the beginning of construction of the K-25 gaseous diffusion facility at Oak Ridge, Tennessee in 1943 (see figure 2.1):

Ground was not broken for the main process buildings until September. The blueprints prepared by Kellex called for a cascade building truly gargantuan in scale. The rough drawings prepared the previous winter had now evolved into a plan for a series of fifty-four contiguous four-story buildings constructed in the shape of a U, almost a half a mile long and more than 1,000 feet in width. The enormous area of the buildings (almost 2,000,000 square feet) and the weight of the process equipment they would contain required an extraordinary amount of earth-moving and unusual techniques in constructing foundations. The conventional method of excavating foundations only under load-bearing walls and columns would have required the design and setting of several thousand columns of many different lengths. As a shortcut, Kellex decided to level the whole area and fill in the low spots with scientifically compacted earth. Since, over the half-mile length of the U, original elevations differed by as much as fifty feet, it was necessary to move almost 3,000,000 cubic yards of earth. The slow job of earth-moving and compacting fill continued in the fall of 1943, and it was not until October 21 that the first 200,000 cubic yards of concrete were poured in the process area. [8e]

Figure 2.1. K-25 gaseous diffusion plant under construction at Oak Ridge, Tennessee, in 1943
The U-shaped building contained thousands of pumps and converters for concentrating uranium-235. Service facilities can be seen in the centre.

Photograph courtesy of Oak Ridge National Laboratory, Oak Ridge, Tennessee.

It is almost incredible that a construction effort of this magnitude was undertaken at a time when the gaseous diffusion process had been demonstrated only on a laboratory scale and before the problem of designing a diffusion barrier suitable for UF_6 had been solved. There were to be no feasibility studies, pilot plants or demonstration facilities. In fact, the barrier problem was still not resolved in June 1944, by which time over one-third of the construction on K-25 had been completed at a cost of $281 million, and cascade stages, without barriers, were already being installed [8f].

The barrier problem was finally solved, and K-25 did contribute some enrichment to the Hiroshima bomb, but the real impact of gaseous diffusion came after the war. With so much capital already invested in the Oak Ridge facility there was little inclination on the part of the US

15

government to begin a new search for alternative and possibly more efficient processes. Then, as other countries, such as the UK and the USSR, accelerated their own nuclear weapon programmes, the proven efficacy and reliability of gaseous diffusion played a crucial role in their own choices of enrichment technology [9]. Both France and China also chose gaseous diffusion for their early nuclear weapon development programmes, and even as late as 1973, at a time when the gas centrifuge was seen by many as a more promising method, France made the conservative decision to use gaseous diffusion in its Eurodif facility at Tricastin [10] (see figure 2.2). On the basis of this historical record gaseous diffusion can be said to have proved itself a highly proliferation-prone process.

Figure 2.2. A view of the modern gaseous diffusion facility (10 800 t SWU/yr) at Tricastin, France

Source: Eurodif S.A., Bagneux, France.

It was only in the 1970s that other enrichment processes began to undermine the pre-eminence of gaseous diffusion. It is interesting to note that each of the modern processes can be shown to have its roots in methods that were tested and found wanting in the early competition with gaseous diffusion. For example, work on gas centrifuges by Beams in the United States [11a] and Harteck and Groth in Germany [12] was intense in the early years of World War II.

The modern use of lasers for selective photoexcitation has its origins in early work on photochemical attempts to separate isotopes [11b]. The ion cyclotron resonance method had its precursor in the magnetron or resonance methods tried in the USA in 1942 [11c], and a large number of chemical-exchange methods were experimented with in the early days. Generally these failed for lack of adequate catalysts.

Great strides in technology and the promise of a large market for enriched uranium have been necessary to reawaken interest in these techniques and others which have been developed since then. Now there are hundreds of ideas on how to separate uranium isotopes, and patents on new ideas are still being issued at a steady rate. From this large field it is possible to identify 10 methods which have either proved themselves viable or are sufficiently promising to be the subjects of intense research and development efforts. These methods are described in quantitative detail in Part Two. In this Part only qualitative comparisons will be used to evaluate the relative susceptibilities to misuse of the various techniques for the production of nuclear explosive material.

III. Comparison of techniques

A number of attempts of varying depth and technical level have already been made to compare various enrichment techniques with regard to their proliferation sensitivity [13a, 14, 15]. They all illustrate the difficulty involved in estimating the effect any technique is likely to have on the proliferation of nuclear weapons. This study has encountered the same difficulty, and even though the technical data are presented in Part Two (see table 6.3, p. 188) in a convenient and readily comparable form, the fact remains that there is no single straightforward and quantitative index which ranks techniques according to their proliferation dangers. This is true even if one ignores the added complications of historical, political and economic factors.

The greatest proliferation danger would be associated with any enrichment process which turned out to be cheap, simple and compact. But none of the existing processes satisfies these criteria, and the prospects that such a process will ever be found are by now almost vanishingly small. The very small percentage of ^{235}U in natural uranium and the very small differences between ^{235}U and ^{238}U atoms strongly suggest that the production of nuclear weapon-grade ^{235}U will always be an expensive and complex process.

However, history has shown that these technical obstacles are not sufficient to prevent the spread of enrichment techniques, a spread which promises to accelerate during the next decade if nuclear power increases its contribution to world energy supply. It also seems intuitively clear that some techniques are more susceptible to various kinds of abuse and misuse than others. It is therefore worth the effort to see whether these intuitive impressions can be made more concrete.

There are two general situations under which the proliferation susceptibility of a technique can be assessed:

1. A country already possesses an enrichment facility capable of producing low-enriched (3–5 per cent) uranium in substantial quantities. How difficult is it to supplement, reprogramme or rearrange such a facility to produce weapon-grade material?

2. A country has no indigenous enrichment capability but decides to construct a small 'dedicated facility', dedicated, that is, to the production of nuclear explosives. Are there some processes for which this task is easier than for others?

Table 2.1 lists the 10 enrichment processes analysed in detail in this study and assigns to them relative ratings with respect to five properties which were introduced in chapter 2 and which seem particularly crucial for proliferation. These ratings are in no sense intended to be quantitative, and the numbers do not have the same absolute or relative meanings in different columns. They are intended only to give a rough high – intermediate – low (1–2–3) estimate for each process. The numbers do not imply fixed quantitative ratios (e.g., the separation factor of the centrifuge is not twice that of gaseous diffusion), and the properties that head the columns are not necessarily independent variables (e.g., the ease of batch recycle is correlated to some extent with equilibrium time).

One way to interpret the ratings is to compare the proportions of 3s to 1s for different processes. However, even this can be misleading since a rating of 3 in only a single category can represent a highly significant barrier to proliferation (e.g., the doubtful feasibility of batch recycling to high enrichments in the AVLIS process). Even the perfect record of 1s attributed to the MLIS process must be interpreted in the light of the uncertain prospects for the development of reliable laser systems which would not be so highly complex and expensive as to put them beyond the reach of any but the richest and most industrially advanced nations.

The qualitative ratings in table 2.1 are based on the quantitative results of Part Two (see table 6.3, p. 188). However, for the purposes of this analysis the following short descriptions of individual processes will serve to explain the assignments made in table 2.1.

1. *Gaseous Diffusion*: The separation factor of a gaseous diffusion stage is limited to a very small value by the small relative mass difference between $^{235}UF_6$ and $^{238}UF_6$. For this reason a gaseous diffusion cascade consists of many stages, and to achieve useful amounts of separative work the stages must be quite large. A typical gaseous diffusion plant covers many hectares of floor space and consumes enormous quantities of electrical power for its pumps and compressors. The equilibrium time and inventory are correspondingly large and for this reason batch recycling to obtain high enrichments is utterly impractical. Since only UF_6 gas is used as the process material, no reflux chemistry is necessary, nor do serious criticality problems arise, even when a highly enriched product is made.

2. *Gas Centrifuge*: Modern centrifuges can be made from new

Table 2.1. Important enrichment technique property ratings according to their contribution to proliferation sensitivity

	Separation factor	Equilibrium time and inventory	Size of dedicated facility	Ease of batch recycle	Reflux chemistry and criticality problems
Gaseous diffusion	3	3	3	3	1
Centrifuge	2	1	1	1	1
Aerodynamic Nozzle	3	1	2	2	1
Helikon	3	1	2	1	1
Chemical Solvent extraction	3	3	3	3	2
Ion exchange	3	3	3	3	2
Laser Molecular (MLIS)	1	1	1	1	1
Atomic (AVLIS)	1	1	2	3	3
Electromagnetic Calutron	1	1	3	2	3
Ion cyclotron resonance	1	1	2	2	3

Rating 1 implies that the factor presents a low barrier to misuse of the technique; rating 3 a significant obstacle to misuse; and rating 2 somewhere in between.

high-strength, lightweight materials, such as carbon- or glass-fibre resins. These allow the centrifuge to spin at extremely high speeds: the outer wall of the centrifuge moving at two or more times the normal speed of sound in air. Magnetic bearings and high vacuum isolation chambers reduce friction on the centrifuge to very low values, so that power consumption and wear-and-tear are kept to a minimum.

The high rotational speeds and sophisticated designs of modern centrifuges give them separative powers much greater than was feasible only 10 years ago. Nowadays a plant capable of making enough highly enriched uranium for several nuclear weapons per year can be built with only a few hundred centrifuges taking up a floor space of only a few thousand square metres, including all supporting equipment. Since the centrifuges work on UF_6 gas at very low pressures, the amount of inventory is very small. This, coupled with the relatively high separation factor of the centrifuges, leads to very short equilibrium times, measured in

19

minutes instead of the weeks used for gaseous diffusion. For these reasons batch recycling is a viable option for a small centrifuge facility. Again, reflux and criticality problems do not arise in the cascade because of the use of low pressure UF_6 gas. However, precautions must be taken in the product-collecting area where the gas is returned to the solid phase.

3. *Aerodynamic*: The jet nozzle process developed in FR Germany and the advanced vortex tube process (or Helikon process) developed in South Africa are two variants of the general class of aerodynamic processes. The working substance is a dilute mixture of UF_6 in hydrogen gas which must be pumped through many stages, each with a separation factor somewhat larger than that for gaseous diffusion but much smaller than that of the centrifuge. So, aerodynamic enrichment plants tend to be intermediate in size, inventory and equilibrium time between gaseous diffusion and centrifuge plants (see figure 2.3). However, certain features of the South African process seem to imply relatively short equilibrium times, so that batch recycling is not ruled out, at least in small facilities.

Figure 2.3. The South African enrichment plant at Valindaba
This facility employs the Helikon aerodynamic process, and as of January 1981 had a capacity of about 6 t SWU/yr.

Photograph courtesy of W.L. Grant, Managing Director, UCOR, Valindaba, Republic of South Africa.

Because of the hydrogen carrier gas, reflux is a bit more complicated for these processes than for the previous two, but suitable solutions seem to have been worked out. Criticality problems are absent in these processes because of the very dilute form of the process gas.

4. *Chemical*: The French solvent extraction and Japanese ion-exchange processes involve chemical-exchange reactions between uranium compounds in different phases which are continuously brought into

20

contact. In the solvent extraction method the compounds are dissolved in aqueous and organic liquids and contacted in very tall 'pulsed columns' similar to those used in the chemical industry. In the ion-exchange process, the phases are an aqueous solution and a finely powdered resin through which the liquid slowly filters. In both processes the chemical exchange is promoted by catalysts.

Chemical-exchange reactions involving different isotopes tend to have extremely small separation factors, especially for heavy elements like uranium. This means that the extraction or exchange columns must consist of many stages, usually measured in the thousands. If the chemical reaction rates are relatively fast and if good mixing can be achieved in each stage the stages can be fairly short, but even in the best of circumstances the columns will be large, implying very large inventories and very long equilibrium times. This suggests that batch recycling for high enrichments would be out of the question in a chemical enrichment facility.

Finally, since the chemicals must be recycled to make either process economical, both involve substantial chemical reflux processes. These are energy-consuming and technically demanding, since they must be accomplished without significant remixing of the uranium isotopes. Criticality is also a problem since the uranium is in liquid solution. However, this can be controlled by the addition of strong neutron-absorbing substances to the solutions, at least at low enrichments. Whether this technique could be applied at arbitrarily high enrichments is not clear.

5. *Laser*: The molecular laser method uses either pure infra-red or a combination of infra-red and ultraviolet laser light to dissociate $^{235}UF_6$ molecules in a supersonic gas jet. The UF_6 is carried in a background gas, presumably an inert gas such as argon or nitrogen. The separation factor for this process can in theory be extremely high and this, coupled with the compact size of a separating element, reduces both inventory and equilibrium time to essentially negligible values. A molecular laser facility capable of producing several bombs per year need be no larger than a small warehouse. In such a facility, batch recycling is no problem whatsoever, and reflux and criticality problems are quite manageable.

The atomic vapour process is considerably more difficult and technologically sophisticated. In this process a uranium vapour must be produced at extremely high temperatures, and the portions of the enrichment element which come into contact with the vapour must be able to withstand its intensely corrosive action. The atomic vapour is much less dense than the gaseous UF_6 used in the molecular process, so an atomic vapour facility must be correspondingly larger to handle the same amount of material. Batch recycling is also considerably more difficult because of a tendency for the efficiency of the AVLIS process to decrease as the percentage of ^{235}U in the feed increases. Finally, the use of solid uranium metal and the need to collect and recycle large quantities of uranium, which condense out of the vapour without being processed, create a significant reflux problem. Criticality is no problem in the vapour state, but it would limit the acceptable size of the collector unit for a highly enriched product. However, this would probably not be a seriously limiting consideration.

It must be made clear that both of the laser enrichment processes are still in the research and development stage and that there are not even any pilot plants operating at the present time. The development of suitable lasers for these processes has presented serious technical problems, for which, according to most of the available evidence, solutions are still far away. However, major efforts to solve them are underway in a disturbingly large number of countries.

6. *Electromagnetic*: The calutron, which gave the world its first highly enriched uranium, is no longer considered a viable process for this purpose. It uses the differential deflection of ionized uranium tetrachloride (UCl_4) molecules to separate isotopes, and if the molecular beams are kept at low intensities, very high separations are possible. However, the rate of production of highly enriched materials is then extremely slow, and a single calutron may take hundreds of years to produce enough material for a single bomb. Any attempt to use hundreds of calutrons, such as was done at Oak Ridge in 1944–45, would be extremely expensive and inefficient both in energy and resources.

The ion cyclotron resonance process uses oscillating electromagnetic fields to extract ^{235}U ions from a uranium plasma. To create this plasma, uranium vapour must be produced and the uranium atoms must then be stripped of an electron. This takes reasonably large amounts of energy, as does the resonance excitation process itself. Also, in order to achieve useful separations the process must be carried out in a highly uniform magnetic field inside a superconducting solenoid. This is a device which can carry electric current with no resistance, but which must be kept at temperatures very close to absolute zero by contact with liquid helium.

All of this leads to a very complex and sophisticated process. Just as for the laser methods, the potential separation factors are large and there is no problem with equilibrium time or inventory. A facility for producing significant quantities of highly enriched material would not be particularly large, and batch recycling should be possible. Reflux and criticality problems would be similar to those for the AVLIS process discussed above.

The plasma process is, if anything, even less well understood and developed than the laser processes. This makes any judgements about its ultimate applicability and proliferation sensitivity highly speculative.

With these brief descriptions as background, it is now possible to compare the various processes with respect to the two situations previously defined.

Conversion of an existing facility

To assume that a commercial enrichment facility already exists in a country is to assume a great deal. First, the process must be workable and reasonably efficient and reliable. It is presumably understood by the people who operate it. It is also clear that any enrichment plant designed for the production of reactor fuel in useful quantities will have an SWU capacity more than ample for the production of substantial numbers of nuclear weapons.

The fundamental question is whether the facility can be operated in such a way as to redirect SWUs from the production of relatively large quantities of 3 per cent material to the production of much smaller quantities of 90 per cent material. As table 5.1 (p. 113) shows, this implies putting over 50 times as many SWUs into each kilogram of product if the tails composition (assay) is kept constant. This ratio can be reduced either by allowing the tails assay to increase, a strategy which increases the feed requirements, or by feeding the cascade with already enriched material, which in effect is a batch recycling process. For the latter the cascade must be totally cleaned out and restocked with a full inventory of partially enriched material. It must then be run without withdrawal of product until the enrichment at the product end reaches the desired value. For processes such as gaseous diffusion or chemical exchange this would be extremely costly and time-consuming, requiring enormous amounts of valuable reactor-grade material and as long as several years of waiting time. On the other hand, for a centrifuge or molecular laser facility the batch recycling process could be quite easy, inexpensive and fast. Indeed, in a centrifuge plant built on the modular concept now standard for such facilities, the conversion would only have to involve a few modules, and most of the plant could continue operating normally. One other process for which batch recycling seems to be feasible is the South African Helikon technique.

Batch recycling is not always necessary. For some kinds of cascades, in particular those which combine large numbers of elements in a series-parallel cascade arrangement, an alternative might be to reconnect the cascade. Instead of having relatively few stages, each containing many parallel elements, the new configuration would have many stages each with a relatively smaller number of elements. In this way the SWUs contributed by each element are devoted more to achieving higher enrichments than to generating large production rates. Such a reconnection of elements can be made more or less difficult by choosing different designs for the original plant. Knowledge of these design choices is an important element in any attempt to apply safeguards to such a facility (see chapter 3).

The processes which seem most sensitive to this kind of abuse are again the centrifuge and MLIS processes and to a lesser extent the aerodynamic processes. Processes for which this form of operation seems difficult or even impossible are the chemical-exchange techniques and gaseous diffusion.

A third possibility for a country with an existing facility is to increase the number of separation stages or to construct a separate, smaller facility of the same type to further enrich the product of the commercial facility. If the process or significant components of it are imported, then, of course, this procedure runs a high risk of detection and possible frustration.

The centrifuge again allows the simplest add-on option, but the aerodynamic processes and gaseous diffusion are also relatively easy to expand once they are built. This might be done by constructing an extra cascade to attach to the top of the commercial cascade. In addition, the commercial cascade could be operated in a nearly total reflux condition, that is, the rate of product extraction can be drastically reduced. This increases the enrichment of the product while reducing its flow rate. However, since much less material is needed for the extra cascade, this is not a great penalty. The added enrichment at the input to the extra cascade reduces the number of stages needed to get to high enrichments, and the small material flows allow the cascade elements to be relatively small.

A large commercial gaseous diffusion cascade capable of producing 3 per cent product in normal operation can produce up to 11 per cent enriched product when operated at nearly total reflux. To convert this material to 50 per cent highly enriched uranium (HEU) would require an additional cascade of some 700 stages operating at optimum reflux, and to get 90 per cent would require 1 400 stages. This is still not a trivial task, and the equilibrium time would still be several months. For an aerodynamic facility the task would be much easier, since the enrichment gain per stage is considerably larger, reducing both the required number of stages and, even more dramatically, the equilibrium time (see table 5.1, p. 113). For chemical exchange additional exchange columns might be added to a commercial facility to produce highly enriched uranium. But, both the long equilibrium time and criticality constraints would make this a rather unattractive option.

Construction of a dedicated facility

A country wishing to build an enrichment plant capable of producing weapon-grade material faces serious problems with any of the methods discussed here. Table 2.1 summarizes many of them, but more must be added to get a full picture of the obstacles facing any new entrant to the enricher's club.

Molecular laser isotope separation, which (according to table 2.1) is the most proliferation-prone process of all, does not yet exist as an operating technique. It still faces severe development problems in connection with the lasers and their associated optical systems, and possibly with other components as well. Research and development work on this process has been going on for a decade or more in many technically sophisticated countries, but so far the process still appears problematical. So, to any nation which now or in the near future will be looking for a suitable process for making highly enriched uranium, the MLIS process would have to be considered as unavailable.

This example highlights the importance of technical complexity in determining the proliferation risks for small, dedicated facilities. Clearly this is far less relevant for countries which already have an enrichment facility of the same kind, but it does apply to countries with one type of facility who wish to construct a different kind, possibly better suited to getting high enrichments.

Aside from MLIS the centrifuge emerges from table 2.1 as the process best suited for small, dedicated facilities. However, the factor of technical complexity is also important in this case. The centrifuge is now a demonstrated process, but many of its manufacturing processes, design parameters and operating characteristics are closely held secrets. The fact that one country, Pakistan, was able to obtain most of this information by espionage is certainly an important object lesson on security problems, but it does not necessarily imply that the world can expect an outbreak of such incidents leading to a rapid spread of centrifuge technology. Even if the information could be obtained by others, the problems of developing the necessary technical and industrial infrastructure and finding the money to pay for it would still remain as it does with any technology. It is true that the secrecy mentioned above applies only to modern, highly efficient centrifuges, and that quite complete information on older, less sophisticated models is freely available [16]. But a dedicated facility constructed with these centrifuges would necessarily be a much larger operation and would still require substantial commitments of money, and engineering and scientific personnel.

The aerodynamic processes seem to strike a balance between efficiency and technical sophistication, which has contributed to their acceptance by South Africa and Brazil. The South African process was developed originally with important help from foreign technical experts [17], but a number of the innovations which have made the Helikon process workable seem to have been developed domestically [18]. The technique does not seem to be as complex as the centrifuge, yet it produces substantially higher enrichment factors than gaseous diffusion. It is quite inefficient in energy consumption, and this may limit its commercial applications to countries with cheap coal or hydroelectricity resources, but this is not likely to be a major concern for a country interested in a nuclear weapon capability.

The most threatening processes of all would seem to be those capable of producing highly enriched product in just a few stages, in particular the laser and plasma processes. These would permit the construction of small, flexible facilities which might be kept secret for long periods of time. But so far there has been a rather direct correlation between large separation factors and high technical complexity. This suggests that, barring an unexpected technological breakthrough, these advanced techniques will remain inaccessible to technologically unsophisticated countries for some time in the future.

The preceding analysis is summarized in table 2.2, where each process has been assigned a threshold indicating its relative susceptibility to the two abuses already described (see above, items 1 and 2 on p. 18). These thresholds represent an attempt to compress a wide variety of factors into a single index, a procedure which is always somewhat subjective. These thresholds must be seen to represent the authors' best judgement of the significance of the technological data presented in Part Two. It is anticipated that other analysts may come to different conclusions, but in any case the classification scheme suggested by table 2.2 seems to provide a useful structure for discussion of proliferation problems. The threshold concept is developed further in the concluding section of this chapter.

Table 2.2. Technological thresholds to proliferation

	Threshold		
Strategy	Low	Intermediate	High
Misuse of existing facility	Centrifuge	Gaseous diffusion	Chemical exchange
	Helikon	Jet nozzle	
	MLIS	AVLIS	
	Plasma	Calutron	
Construction of dedicated facility		Centrifuge	Gaseous diffusion
		Helikon	Chemical exchange
		MLIS	AVLIS
			Plasma
			Jet nozzle
			Calutron

This assessment can be concluded with the following general remarks. First, it is clear that there is a strong correlation between the potential usefulness of an enrichment process for commercial applications and its potential for military applications. A process which allows the construction of small, energy-efficient, modular facilities has obvious commercial advantages as well as serious proliferation implications. The gas centrifuge fits this description closely and must therefore be seen as a particularly sensitive technique.

One can be certain that if the demand for enrichment services rises in the future, and if the premium placed on nationally independent nuclear fuel cycles remains high, then it is probably inevitable that the centrifuge process, or one which possesses similar commercially attractive properties, will continue to spread.

A partial answer to this problem may lie in the creation of multinational or international enrichment centres which guarantee supplies of low-enriched fuel to their customers (see chapter 4). But if these centres also rely on sensitive techniques, then the possibility of diffusion of knowledge about the technique must be accepted. If the staff of a multinational facility is also multinational, then it is always possible that security will be breached and sensitive information will be acquired by individual countries.

To counter this it can be argued that multinational facilities should employ technology which benefits from economies of scale and is much less likely to spread. It is in this sense that the chemical-exchange or gaseous diffusion processes might be useful in controlling proliferation. Even if the centrifuge should still prove to be more economical on these large scales, which seems quite likely, the added cost of the other processes could be justified as a price the nations involved are willing to pay to prevent proliferation.

IV. Incentives and motivations

The development of uranium enrichment techniques has always been closely related to the overall development of nuclear technology, both military and civilian. Different countries have had different motivations for acquiring enrichment capabilities, and even within individual countries these motivations have varied over time. For some countries nuclear energy development is considered important for the stimulation of industrial and technological innovations. For others an interest in uranium enrichment is part of a resource development strategy or the result of a desire for energy independence.

To this list of motivations must be added the desire for greater national security or international power and prestige. Because many countries see military power as essential to these goals, and because the possession of nuclear weapons is still seen by many to contribute to military power, the military incentive for the development of nuclear energy, and in particular uranium enrichment, remains very much alive. And, as the previous analysis has demonstrated, all enrichment processes possess features which make them more or less susceptible to use for military purposes.

The simultaneous existence of military and civilian applications for nuclear technology and the possibility that both peaceful and aggressive motivations can be at work in forming national nuclear policies makes the analysis of such policies extremely difficult. Our approach to this problem, as outlined briefly in the introduction to this chapter, is to attempt to separate objective factors encouraging the spread of enrichment technology from the less tangible motivations which may push certain states to consider acquiring such technology. This analysis will follow after a brief historical review of the development of the uranium enrichment industry.

Historical development

First phase: military motivations 1945–1960

The birth of the uranium enrichment industry has already been described in section II of this chapter. Gaseous diffusion emerged from the short and intense US development effort as the only technique which had demonstrated its ability to enrich uranium reliably to whatever degree of enrichment was desired. The process was almost grotesquely inefficient and expensive. But it worked; and given the powerful incentives that existed in several countries to acquire nuclear weapons, gaseous diffusion became the process of choice for all early enrichment facilities.

After the war the UK started its own nuclear weapon programme based on gaseous diffusion [19, 20]. The enrichment plant was constructed at Capenhurst and came into operation between 1954 and 1957. Also in 1945 the French started a nuclear programme, and although its early leader Frédéric Joliot-Curie strongly opposed any production of nuclear weapons [21], subsequent leaders took a different course. By the early 1950s France had a growing nuclear weapon programme [22]. The original French programme was based on natural uranium and plutonium, but in 1960 construction was begun on a gaseous diffusion plant at Pierrelatte. France made the decision to develop a national enrichment capability only after several attempts at international collaboration had failed [23a, 24]. These attempts will be discussed further below. The Pierrelatte plant began operation in 1964 and was completed in 1967.

The USSR wasted no time in launching its own nuclear weapon programme during the latter stages of World War II. Gaseous diffusion was chosen as the enrichment process, and a Soviet facility was producing enriched uranium by the early 1950s [9]. In the context of the Soviet–Chinese nuclear collaboration of the 1950s, the USSR assisted in the construction of an enrichment plant within the People's Republic of China. After the break between the USSR and China in 1959 the Soviet Union stopped its nuclear assistance programme, but the Chinese succeeded in finishing their enrichment facility, which is generally assumed to employ gaseous diffusion.

All of these enrichment programmes were motivated primarily by a desire for nuclear weapons. It is significant that in this early stage of development all countries chose to adopt the extremely expensive and highly inefficient gaseous diffusion technology. Although no clear statements of the motivations for this choice are available, it seems reasonable to assume that the urgency of these countries' desire for nuclear weapons led them to choose technology they knew would work rather than engage in a long research and development programme to perfect possibly more economical and efficient processes.

The reader should be reminded at this point that highly enriched uranium is not the only material from which nuclear weapons can be made. In fact, all the nations which developed their nuclear weapons in the 1940s, 50s and 60s also acquired plutonium production facilities and tested plutonium weapons. But all of these countries also acquired uranium enrichment capabilities, probably for several reasons, but certainly in part because of their desire to build hydrogen bombs.

It is also important to recognize that the classification of diffusion barrier and compressor techniques by the USA had little effect on the ability of other nations to develop these techniques independently. This suggests that the design of usable barriers and other components is well within the capability of any nation with a reasonably advanced technology base, and that the high costs of a gaseous diffusion plant are a much greater obstacle to proliferation than any secrets of its design or operation.

Second phase: commercial motivations 1950—1970

Even in the early stages of nuclear development there were commercial motivations. However, these were generally subordinate to military considerations. Only after the announcement of the Eisenhower 'Atoms for Peace' programme in 1953 did the commercial potential of nuclear energy begin to emerge as a prominent factor in development. The prospect of cheap electrical energy and its assumed stimulation of industrial development in both developed and Third World countries led to a rapidly growing research and development effort all over the world. Although most of this optimism has been dissipated by subsequent events, it was a very important element in creating the technological momentum which exists today.

Both the USA and the USSR stimulated the development of nuclear technology in other countries through agreements on co-operation in this field. By 1955 the USA already had 22 bilateral contracts, while the USSR had 7 with other Socialist states. However, both nations guarded their predominant positions, mainly by keeping sensitive know-how secret even as they provided special materials and components. Many analysts saw this as a mechanism for maintaining some control over nuclear weapon proliferation [25].

Uranium enrichment played an important role in these strategies, especially in the West. Light-water reactors, which use low-enriched uranium for fuel, became the dominant reactors in the USA and were strongly promoted for export. The United States could supply the fuel for these reactors at relatively low prices because of the large capacity of its existing enrichment plants, only a fraction of which was needed to meet the military demand for weapons after the early 1950s. These low fuel prices not only made US reactors more attractive, they also made it economically unfavourable for other countries to make the large investments necessary to develop their own enrichment capabilities. In this way the USA was able to delay for some time the proliferation of enrichment capabilities. This unilateral mechanism for proliferation control is analysed further in chapter 4.

The supply-and-demand picture for uranium enrichment as it appeared at the end of the 1960s is illustrated in tables 2.3 and 2.4. Table 2.3 gives existing enrichment capacities at that time and translates them into the nuclear electric capacity (based on light-water reactors) which they would support[1]. Table 2.4 shows some projections for the growth of nuclear electric-generating capacities which were widely accepted as valid in 1970. If the two tables are compared, under the assumption that light-water reactors would dominate the power reactor market, it is clear that the demand for enrichment services would have exceeded existing supply by the late 1970s. In addition, the USA would have had a virtual monopoly on enrichment services outside the Socialist Bloc.

Of course, this situation had been anticipated long before 1970, and during the 1960s research and development continued on alternative enrichment techniques, in particular the gas centrifuge and jet nozzle processes. Also in the 1960s the future prospects for a greatly expanded enrichment market looked promising enough for a number of nations to begin serious planning of new enrichment facilities.

Having suffered failure in an earlier attempt at European co-operation in enrichment (the failure that led to the French plant at Pierrelatte), a number of countries tried again for such a collaboration in 1966 (see chapter 8). This proposal was made by FR Germany and Italy, who suggested that the UK and France contribute their technological know-how to the project [23b]. However, neither country was willing to do this, and in turn the West Germans rejected a British proposal that FR Germany invest in an expansion of the Capenhurst facility, which would remain entirely under British control.

---

[1]   A new unit of enrichment capacity, t SWU, is introduced in table 2.3. This stands for tonne SWU and represents 1 000 of the SWU which were defined earlier in chapter 1 (see p. 10). The latter are often called kg SWU. The unit of electrical capacity, GW(e), is called gigawatts-electric. One GW(e) is one million kilowatts of electrical generating capacity, about the capacity of a modern nuclear power plant.

Table 2.3. Gaseous diffusion plants built for military purposes

Country	Capacity (10^3 t SWU/yr)	Production sufficient for Number of bombs/yr[a]	Production sufficient for Electricity generation (GW(e))[b]
USA	17.2	4 000	160
USSR	7–10	2 000	80
UK	0.4	100	3.6
France	0.3–0.6	75–150	2.7–5.4
China	0.18	45	1.6

[a] Very rough estimates based on about 20 kg of 95 per cent enriched uranium per bomb at an operating tails assay at the enrichment plant of about 0.25 per cent.
[b] Rough estimate based on a need for 110 t SWU/GW(e)/yr. For more details see appendix 8B.

Table 2.4. 1970 projections of nuclear growth (in GW(e))

Projection for the region	Projection for the year 1975	Projection for the year 1980	Projection for the year 1985	Reference
USA	59	150	300	US–AEC [26]
European Community	28.6	74.6	148.4	OECD [27]
World (non-CPE)[a]	118	300	610	OECD [27]

[a] Centrally planned economy.

The first successful European collaboration was achieved when the Treaty of Almelo was signed by FR Germany, the UK and the Netherlands in 1970, creating the Uranium Enrichment Company (Urenco)[28]. The Treaty, which came into force in 1971, provided for the construction of three 25 t SWU/yr pilot plants based on gas centrifuge technology. One of the plants was to be located at Capenhurst and to be under British control, and the other two were to be built at Almelo in the Netherlands, with one under Dutch and the other under West German control.

The Treaty provided for a further development of gas centrifuge technology, aiming at ultimate integration of the programmes in an optimized centrifuge concept and the creation of production facilities on a commercial scale. The pilot plants in Almelo were the first enrichment facilities on the soil of a non-nuclear weapon state. They were intended as the nucleus for a larger facility, producing low-enriched nuclear fuel both for the nuclear programmes of the three participating countries and for the world market.

31

Third phase: expansion and stagnation 1970–1980

As the 1970s began, all signs seemed to point towards a rapidly expanding market for enrichment services. New ideas for enrichment techniques were being explored, and some old ones were being re-examined in light of the new economic and technological situation. Research and development programmes were accelerating in many countries, and in the USA the Nixon Administration became convinced that the enrichment business should be turned over to private enterprise. This decision encouraged a number of private companies in the USA to undertake their own extensive R&D programmes.

In Western Europe the success of the Urenco collaboration was followed quickly by another successful multinational agreement. This was the creation of Eurodif, a co-operative effort involving France, Italy, Belgium, Spain and Iran to build a large (10 800 t SWU/yr) gaseous diffusion plant at Tricastin. The plant was under French control, but capital and technology were supplied by all of the participating countries. These countries also provided an assured market for 90 per cent of the plant's capacity through long-term contracts for enrichment services.

The two threats to a greatly expanded world nuclear economy dependent on enriched uranium seemed to be the political one of weapon proliferation and the technological one of the plutonium-based breeder reactor. However, with the signing of the Non-Proliferation Treaty in 1968 nuclear promoters were able to rationalize that the proliferation problem was on its way towards solution; and most projections for the breeder reactor saw no significant impact on the LWR market for at least 20 years. Other fuel cycles, such as natural uranium or thorium–uranium-233 were seen as possibly of interest to a few specific countries, but not as a serious threat to the dominance of LWRs.

In this optimistic context there emerged two categories of motivation for entering the enrichment market.

1. Strictly commercial motivations: countries with large uranium deposits or an abundance of cheap electric power (e.g., hydroelectric or coal resources which would enable them to supply enrichment services at relatively low costs) saw a chance to exploit these resources for substantial profit.

2. Energy independence: the US and Soviet monopoly on enrichment services had already become a major irritant in world politics, but it was the oil embargo of 1973 which made the virtues of energy independence clear to all countries, especially to Western Europe and Japan. In this context a national enrichment capability was seen as a desirable component of an independent nuclear fuel cycle.

The rapidly rising costs of energy also fed back into the enrichment R&D effort, putting a new premium on more energy-efficient enrichment technology. The centrifuge rapidly became much more competitive with gaseous diffusion, and incentives increased for the development of laser or chemical-exchange techniques which also promised much lower energy costs per SWU.

The optimism of the early 1970s was short-lived. By the end of the decade it was clear that the enrichment market would not expand at anything like the rates predicted in table 2.4, and that commitments made in the late 1960s and early 1970s had produced a substantial overcapacity of enrichment services. By 1980 most of the countries which had projected greatly expanded nuclear energy programmes had either drastically cut or entirely abandoned these programmes. Public opposition to nuclear power focused on problems of safety, waste disposal, civil liberties and weapon proliferation, especially after the Indian nuclear explosion in 1974. To this growing political opposition were added shortages of investment capital and an increasing awareness that the real economic costs of nuclear energy had been severely underestimated. By the time of the Three Mile Island accident in March 1979 these problems had combined to make nuclear energy both unprofitable and unpopular. The reactions of various groups to the accident serve as an excellent symbol of how drastic had been the shift in public, business and government attitudes towards nuclear energy in the 1970s.

The present situation

The uranium enrichment picture in the early 1980s is characterized by a substantial excess of supply over demand, a multiplication in the number of workable enrichment techniques, a still intense interest in national energy independence, sustained tension in several regions of the world, and a continuing anticipation of energy shortages in the latter part of the decade. Interactions among all these factors have produced a highly unstable situation, and the dangers of nuclear weapon proliferation are at least as great now as they have been at any time in the past.

The combination of large research, development and capital investments in new enrichment techniques and the drying up of the demand for additional enrichment services has led to intense competition among suppliers to capture larger shares of the market which does exist. One dangerous element of this competition could be a willingness to relax requirements for safeguards and other non-proliferation requirements in order to gain an edge over competitors. Tensions created by this tendency have been severe among Western suppliers of enrichment services and technology, and various efforts to control this competition have been made. These are discussed in chapter 3.

This picture of the world enrichment situation provides the background for the following analysis of national enrichment attitudes and policies. A more detailed summary of the enrichment status of individual states is presented in chapter 8. Also contained in that chapter are the data supporting the following assessments of the situational thresholds and motivations of various nations.

Situational thresholds

The definition of situational thresholds is based on the observation that it is easier for some nations than for others to use a uranium enrichment capability to produce nuclear weapons. For example, if a nation has an enrichment facility on its territory and under exclusive national control, then the threshold will be lower than if it imports its low-enriched uranium or shares control in a facility with other nations.

An extremely important situational threshold is created by the participation of the country in the Non-Proliferation Treaty and the application of international safeguards to its facilities. Such participation significantly raises the proliferation threshold in that country.

A third situational threshold is associated with the level of international tension and/or hostility experienced by a state. Assuming that in most non-nuclear weapon states there are significant domestic political obstacles to acquiring nuclear weapons, these obstacles can be lowered considerably by the presence of hostile neighbours or other threats to national security. Such situations tend to encourage those political elements who see some value in a nuclear 'deterrent', and create the impression that such an option is necessary.

The three situational thresholds mentioned above are similar to technological thresholds in that they derive entirely from the objective situation of a country and do not depend on judgements about the peacefulness or aggressiveness of a nation's motives. Taken together, the technological and situational thresholds can be thought of as defining structural thresholds to proliferation. In principle, all states involved in uranium enrichment can be classified according to the heights of these

Table 2.5. A survey of the enrichment situation in terms of thresholds to proliferation through diversion

Inherent situational threshold	Inherent technological threshold		
	Low	Intermediate	High
Low	South Africa (AE) Pakistan (GC) India[a] (GC[?], L[?])	Israel (L)	
Intermediate	FR Germany (GC) Netherlands (GC)	Brazil (AE)	Belgium (GD) Italy (GD) Iran (GD) Spain (GD)
High	Japan (GC,GD,L,CE) Australia (GC,L,CE,PL)		

AE = Aerodynamic. CE = Chemical exchange. GC = Gas centrifuge. GD = Gaseous diffusion. L = Laser. PL = Plasma.
[a] Only occasional reports have appeared with respect to Indian enrichment activities (see chapter 8, pp. 233–35).

thresholds, and this classification can be used to obtain a picture of the overall threat of proliferation posed by enrichment activities. Such a classification is offered in table 2.5. It includes the countries with known enrichment facilities and/or R&D programmes, as listed in table 8.2 (p. 228), excluding the five nuclear weapon states (as defined by the NPT). The thresholds are set at three levels — low, intermediate and high. The thresholds include both options for diversion outlined in section III of this chapter: the conversion or reprogramming of an existing facility and the construction of a dedicated facility. Only a minority of the countries involved have their own operating facilities.

Looking first at technological thresholds it is clear that a trend exists for the development and use of low threshold techniques, especially the gas centrifuge and laser. But there is an important distinction between these two processes: the centrifuge is a proven, essentially mature industrial process, while the various laser enrichment proposals remain in the relatively early stages of research and development. For this reason, laser enrichment methods are given a relatively higher technological threshold, even though they will have to be considered much more dangerous once their practicality is demonstrated. Available evidence can establish only that Israel is engaged in R&D on laser enrichment, so Israel has been given a relatively higher technological threshold. On the other hand, South Africa's Helikon technology would represent at least an intermediate threshold to any country attempting to build it from the beginning. However, once an operating facility and its industrial infrastructure exist, as they do now in South Africa, the technology presents a rather low threshold to modification or reprogramming for obtaining high enrichments. These two examples illustrate the essentially static nature of the ratings in table 2.5 and their susceptibility to change as technology evolves. With respect to India, it must be kept in mind that only occasional reports have appeared concerning its enrichment activities, and that the current status of these efforts is unknown.

With regard to the situational categories, Israel, India, Pakistan and South Africa have been given low thresholds. All four countries are located in actual conflict regions, none of them is party to the NPT or has full-scope safeguards, and their enrichment activities are national programmes. FR Germany, the Netherlands, Belgium and Italy are placed in the intermediate threshold category. These countries are NPT members and thus International Atomic Energy Agency (IAEA) safeguarded, and they are all participants in multinational enrichment activities. However, they are located in a potential conflict area and are members of one of the major military alliances with US-supplied nuclear weapons on their territories. The North Atlantic Treaty Organization (NATO) strategy includes the participation of their armies in nuclear warfare within the context of the alliance. As for FR Germany, which considers itself part of a divided nation, it has been suggested that a dissolution or serious weakening of the NATO alliance might be interpreted as a threat to supreme West German interests and cause a reconsideration of FR Germany's commitment to the NPT [29]. For other West European countries, which have grown accustomed to accepting nuclear weapons as part of their national security policies, such an event could also represent a critical development.

Brazil has been given an intermediate situational threshold mainly because of her insistence on keeping open the option to test nuclear explosive devices, albeit for peaceful purposes. Spain is not a member of the NPT, and Iran is located in a conflict region and at war with Iraq. For the Eurodif countries Belgium, Italy, Spain and Iran, it should be kept in mind that the enrichment facility they use is not located on their territories. For this reason, Iran has not been given a low threshold despite its location in a politically unstable area and the war with Iraq. Japan and Australia are categorized in the high threshold category because they are members of the NPT and thus internationally safeguarded, and because they are located in relatively stable political environments with no manifest international conflicts.

It is clear that if table 2.5 were to be compared with a similar one reflecting the situation 10 years earlier the result would show a trend towards lower technological and situational thresholds. Six of the 13 programmes involved are characterized by either a low and intermediate threshold category or by a low threshold score on both categories; and the situation could become substantially worse if laser enrichment techniques turn out to be successful. Thus, from a structural point of view, developments in the uranium enrichment field must be characterized as increasing the risks of proliferation of nuclear weapon capabilities.

Motivations for an enrichment capability

As important as objective structural factors are, they are not sufficient for a full assessment of the proliferation risks associated with uranium enrichment. Some insight is also necessary into the motivations of states who are interested in developing or acquiring an enrichment capability.

The analysis of motivations begins with the countries listed in table 2.5, all of whom have earned a place in the table by actively pursuing research and development on enrichment techniques or by the actual acquisition of or participation in an operating enrichment facility. The next step is then to enquire whether non-military motivations for such interest or participation can be found in the economic, technological and political status of the country. Examples of such motivations might be the combination of a dependence on nuclear electric power and a desire to acquire an indigenous fuel production capability. Such motivations would be quite understandable given the uncertain future of fossil fuel supplies and the growing general awareness of the benefits of energy independence. However, for a country without a significant nuclear energy programme this cannot be the motivation for acquiring an enrichment capability. Even for countries with several nuclear reactors this motivation becomes less convincing when the current over-supply of world enrichment capacity and the rich variety of potential suppliers is taken into account.

Another legitimate motivation for the acquisition of an enrichment capability might be the presence in a country of known uranium deposits. By enriching this material before exporting it a nation could substantially increase the economic benefits derived from such resources, assuming that sufficient demand existed for the enriched product.

One more possible source of interest in such advanced technology as lasers, plasmas or centrifuges might simply be part of a nation's general desire to remain competitive with other nations across a broad spectrum of technological options. This motivation seems less convincing than the previous two, even for a technologically advanced country. It is far less convincing for relatively backward countries in which research and development on powerful infra-red lasers or high-speed centrifuges diverts scarce resources and scientific talent from urgent developmental needs. Therefore, it is worthwhile to determine whether the research and development interests of a country are in fact consistent with its general technological level, or whether some distortion of priorities is present. Such a distortion may raise suspicion that the country is looking for a short-cut to a nuclear weapon capability.

A final clue to the motivations of a state in its pursuit of an enrichment capability is its general behaviour towards the non-proliferation issue. This behaviour, as it is expressed through adherence or non-adherence to the NPT, acceptance or non-acceptance of safeguards, and official statements of nuclear policy, can be used to assess to what degree motives for nuclear weapon acquisition do or do not exist.

It needs to be emphasized that the non-proliferation aspects used as indicators for the situational thresholds are not identical to those indicating non-proliferation attitudes. The former refer to the factual situation with regard to the NPT, safeguards, and so on, while the latter constitute an interpretation of motives. Whereas the former are relatively objective, the latter are more subjective. There is a strong tendency, at least among the authors of this study, to assume that adherence to the NPT and acceptance of effective safeguards is a particularly convincing way for a nation to back up its professed lack of interest in acquiring nuclear weapons.

Using data from Part Three it is possible to assess each of the countries listed in table 2.5 with respect to the above criteria. For each country with a demonstrated interest in uranium enrichment it can be asked whether such an interest is consistent with any of the three motivations discussed above. Consistency is indicated in tables 2.6a and 2.6b by a '+' symbol while inconsistency is indicated by an 'o'. Then, as an additional measure the non-proliferation attitude of the country is similarly assessed, a '+' representing a positive attitude (signatory of the NPT, acceptance of safeguards, etc.) and an 'o' representing a negative attitude. Note that the countries listed in table 2.6a have in common at least one low structural

Table 2.6a. Possible motivations of relatively low threshold countries*

	Pakistan	India[a]	Israel	South Africa	FR Germany	Netherlands
Energy independence	o	o	o	+	+	o
Resources development	o	o	o	+	o	o
Industrial/technological development	o	+/o	+	+	+	+
Proliferation attitude	o	o	o	o	+	+

* Categories low-low, low-intermediate, intermediate-low of table 2.5.
[a] Only occasional reports have appeared with respect to Indian enrichment activities (see chapter 8, pp. 233–35).

Table 2.6b. Possible motivations of relatively high threshold countries*

	Brazil	Japan	Australia	Iran	Italy	Spain	Belgium
Energy independence	+/o	+	o	+/o	+	+	+
Resources development	+	o	+	o	o	o	o
Industrial/technological development	+	+	+	o	+/o	o	o
Proliferation attitude	+/o	+	+	+	+	o	+

* At least one high threshold or both intermediate.

threshold in table 2.5, while those in table 2.6b are all either intermediate or high threshold countries. Table 2.6a shows that of the six low threshold countries, four must be categorized as having weak non-proliferation attitudes. Israel and South Africa are generally considered to have nuclear weapons in an advanced state of development. Israel is estimated to have access to between 10 and 20 nuclear weapons at very short notice, and strong evidence exists that South Africa has prepared for and possibly carried out at least one successful nuclear weapon test (see discussion on p. 77). Moreover, neither country is party to the NPT and the South African enrichment facility is not safeguarded. Neither Pakistan, India, nor Israel possess motives for energy independence based on a nuclear power programme using enriched uranium or the development of large national uranium resources. For South Africa such motivations could certainly be present, but the reader should be reminded that such a consistency does not imply the absence of intentions for nuclear weapon applications. Thus, given their low threshold status and at best ambiguous motives, these four countries should be considered serious risks for proliferation. The non-proliferation attitudes of FR Germany and the Netherlands can be considered positive, and the West German nuclear power programme is heavily dependent on enriched fuel.

Table 2.6b shows that for a majority of the high threshold countries motives of energy independence through nuclear power programmes and resources development can be considered consistent and are generally accompanied by positive non-proliferation attitudes. This holds especially for the European Eurodif countries and Japan on the issue of energy independence, and for Australia and Brazil on the issue of the development of indigenous uranium resources. Italy has been assigned a positive rating for its industrial/technological development motivation on the basis of its substantial contributions of advanced technology to the Eurodif facility. As discussed in the country studies of Part Three, the situation of Iran is rather unclear because of the retardation of its nuclear power programme. The Brazilian non-proliferation attitude must be considered ambiguous, since Brazil has not signed the NPT but has had a positive attitude towards the Tlatelolco Treaty, except for explicitly leaving open the option of underground testing of nuclear explosive devices similar to nuclear weapons.

V. Summary

In summary, the present situation can be described as follows.

1. Enrichment activities with low thresholds to diversion are being conducted in at least four countries where motives for doing this can also be considered as implying a high risk for proliferation of nuclear weapons.

2. For a majority of the advanced industrialized countries motives of energy independence can be considered consistent with the development of enrichment capabilities and correlate with positive non-proliferation attitudes and high structural thresholds for diversion.

3. Given existing trends, the risk of nuclear weapon proliferation through the uranium enrichment route is definitely increasing. This is true because more countries with ambiguous proliferation policies may soon acquire their own enrichment capability and may find themselves with increasingly ready access to poorly safeguarded uranium at the enrichment plants.

It is clear from this summary that the evolving enrichment industry poses substantial problems for the control of nuclear weapon proliferation. Several possible approaches to the solution of these problems are analysed in the next chapter.

Chapter 3.   Options for control

I. Introduction

Uranium enrichment has always played an important role in nuclear weapon proliferation. Every currently acknowledged nuclear weapon state has developed an enrichment capability as part of its weapon programme, and every non-weapon state which possesses an enrichment facility is considerably closer to a weapon capability than it would be without one. As chapter 2 has shown, the degree of proliferation risk varies depending on the type of technology involved and the political situation of the state, but the overall effect of the spread of enrichment technology is to increase substantially the risk of weapon proliferation.

Efforts to control the spread of nuclear weapons are as old as the weapons themselves, and these efforts have been very much a part of the development of the uranium enrichment industry. These efforts have taken many forms, ranging from technical to diplomatic and from unilateral to international. A plausible case can be made that without these efforts proliferation would most likely have proceeded further than it has, but once this is conceded, opinions differ markedly on which methods have been most effective and on where further efforts are likely to be most productive.

It is the purpose of this chapter to describe and analyse the various attempts to control proliferation in terms of their applicability to uranium enrichment. Non-proliferation efforts of the past have been considerably less concerned with enrichment than with other aspects of nuclear technology, in particular plutonium production and distribution. However, given the technological advances of the past decade it is now necessary to examine the applicability of existing and proposed control mechanisms to the enrichment industry much more thoroughly than has previously been done.

With this objective in mind the chapter has been divided into two major sections. Section II takes up the technical problem of applying safeguards to enrichment facilities, and section III evaluates the non-technical, or institutional mechanisms which have been attempted in the past or proposed for the future.

This chapter is largely analytical and assumes some knowledge by the reader of past non-proliferation efforts and the evolution of the uranium enrichment industry. Readers unfamiliar with these areas are referred to Part Three where both a historical review of non-proliferation measures and a survey of the current and projected enrichment industry are given. All of the data on which the analysis of this chapter is based can be found in Part Three.

II. Technical control mechanisms: safeguards

Basic concepts

A major component of the international effort to prevent the spread of nuclear weapons has been the application of safeguards. This word was first used in November of 1945, only three months after the atomic bombings of Japan, in the "Three Nation Agreed Declaration" on international atomic energy policy signed by Canada, the UK and the USA [30a]. Since that time the concept of safeguards has undergone a continual evolution, and it remains a topic of intense interest and sometimes controversy.

There is no single universally accepted definition of safeguards. An idea of the spectrum spanned by this concept can be obtained from two very different statements of the objectives of a safeguards system. One particularly strong statement gives the following four functions as necessary to prevent nuclear theft and violence [4c]: prevention of theft; detection of theft; recovery of stolen material; and response to threats of violence. Although this list refers to theft of nuclear materials, presumably by so-called subnational groups, it could as well apply to diversions and misuse of these materials by nations. Just such a comprehensive safeguards function was intended to be incorporated into the international atomic energy authority proposed in the Baruch Plan [4d]. It is interesting to recall how difficult the creation of an effective safeguards system seemed to the experts at that time. The Acheson–Lilienthal proposal stated that "for a diffusion plant operated under national auspices, to offer any real hope of guarding against diversion, 300 inspectors would be required" [31a].

At the other end of the spectrum is the objective of the current safeguards system of the IAEA: "the timely detection of diversion of significant quantities of nuclear material . . . , and deterrence of such diversion by the risk of early detection" [6b]. Note that the objectives of prevention, recovery and response are not included. The best that has so far been achievable has been a system designed to deter diversions by threat of early detection. This is hardly surprising, since "Arming an international agency with the capability to prevent governmentally authorized nuclear diversions would be politically revolutionary" [4e]. This statement could apply as well to the recovery and response functions.

This book is not the place for a thorough review of the IAEA safeguards system. Several excellent reviews already exist, and the reader is referred to these for a more comprehensive treatment [4, 15, 30, 32, 33]. Here, since the focus is on uranium enrichment, it will have to suffice to show how well the current safeguards regime can be expected to control this particular route to nuclear weapon proliferation. Another limitation on this analysis will be to confine it to safeguards measures applied at the site of the enrichment plant. No consideration will be given to the problem of diversion of enriched uranium after it leaves the plant. Even with these limitations the analysis will be difficult, since the IAEA has had very little experience in safeguarding enrichment facilities. The first statement of the Agency's safeguards system, Information Circular (INFCIRC)/66/Rev 2, rather noticeably excluded enrichment plants from the scope of the document. The only specific mention of enrichment facilities is to make clear that they are *not* to be considered "conversion plants", that is, plants which "improve . . . nuclear material . . . by changing its chemical or physical form so as to facilitate further use or processing" [34].

A more recent document, INFCIRC/153 (the 'Blue Book'), does mention enrichment plants as part of the full-scope safeguards to be applied to the facilities of signatories of the Nuclear Non-Proliferation Treaty. But it is only in the last few years that the IAEA has begun to take seriously the task of designing and implementing safeguards systems for enrichment facilities. A measure of the smallness of this beginning can be gained by noting that only three enrichment facilities are listed as being under IAEA safeguards, one in Japan and two pilot plants at Almelo in the Netherlands [3b]. Closer examination shows that the Japanese facility is a small research and development effort, and that the IAEA safeguards agreement at the present commercial Almelo plant is still under negotiation among Urenco, the European Atomic Energy Community (Euratom) and the IAEA.

Given this sparse record and the complete secrecy under which the negotiations concerning the Almelo plant are taking place, this account will necessarily be rather speculative. Nevertheless, it should be possible to get a reasonably good idea of what an enrichment plant safeguards system might look like by combining the past record of IAEA safeguards efforts at other facilities with the special characteristics of enrichment plants.

Enrichment plants can be treated in one of three ways with respect to safeguards:

1. As part of a full-scope safeguards agreement under the NPT and administered according to the guidelines laid down in INFCIRC/153. An example would be the Netherlands' Almelo facilities.

2. As part of a specific safeguards agreement between a state or states operating the facility and the IAEA. In this case the guidelines come from INFCIRC/66/Rev 2 and apply only to the facilities specifically mentioned in the agreement. For example, Brazil, which is neither party to the NPT nor to the Treaty of Tlatelolco (which it has ratified but by which it is not yet bound; see chapter 8), has agreed to place under safeguards the facilities acquired from FR Germany as part of their major transaction, involving an entire nuclear fuel cycle, including the future facilities derived from them. This agreement includes an enrichment plant based on the jet nozzle process [35].

3. No international safeguards at all. Here there are far too many examples, including, of course, the enrichment facilities of all nuclear weapon states as well as those of Pakistan and South Africa.

Case three will be considered below as part of a general assessment of the limitations of safeguards as a mechanism for controlling proliferation. For now it will be assumed that the facility in question is under one of the first two safeguards categories. There are differences between these two, largely in the degree to which certain requirements are made explicit [30b]. But, most of these differences are too subtle to affect the kind of quick review intended here. Both INFCIRC/66/Rev 2 and INFCIRC/153 are so general that virtually all of the details of the safeguards agreements are left to negotiations between the IAEA and the state concerned.

Safeguards objectives

As was noted above, the objectives of an IAEA-administered safeguards system are the timely detection of diversions of significant quantities of nuclear material. A significant quantity is defined by the IAEA to be "The approximate quantity of nuclear material in respect of which, taking into account any conversion process involved, the possibility of manufacturing a nuclear explosive device cannot be excluded" [36a]. The quantities given which are relevant to an enrichment plant are 25 kg of ^{235}U in uranium enriched to 20 per cent or more, 'highly enriched uranium' (HEU), and 75 kg of ^{235}U in material enriched to lower values, 'low-enriched uranium' (LEU). The amount of ^{235}U is larger in the second case because of the assumed difficulty of the conversion process involved in further enriching it to weapon-grade levels.

The 25 kg value for ^{235}U in HEU is based on an estimate of 25 kg as the minimum critical mass of ^{235}U (90–95 per cent enriched) needed to make a nuclear weapon. It is interesting to compare this with the value of 15 kg in table 1.1 (p. 5), and to ask whether the difference is significant. The lower estimate is based on the assumption that a good neutron reflector is used, but it apparently does not depend on the assumption that a sophisticated implosion system would be used to trigger the weapon. With such implosion systems even smaller amounts of very highly enriched ^{235}U are sufficient to make a weapon [4f]. It seems, therefore, that the IAEA's determination of what constitutes a significant quantity of ^{235}U may be based on some possibly unwarranted assumptions about the low technical sophistication of weapon builders who might use the diverted material.

The choice of 75 kg of ^{235}U in LEU is also questionable. The assumption that the conversion of this material to HEU would be very difficult may be much less realistic now than it was when only gaseous diffusion was available as an enrichment process. For example, if the 75 kg of ^{235}U were contained in 3 per cent reactor fuel, this would represent 2 500 kg of uranium[1]. But it can be shown that the total inventory of a small (60 t SWU/yr) centrifuge plant designed for the production of 90 per cent product from 3 per cent feed would be only about 1 kg[2]. The equilibrium time of the plant would be 44 minutes. Such a plant would produce weapon-grade uranium at the rate of 3.1 kg/day, and it would take only about 18 days to process the 2 500 kg of diverted feed material, turning it into enough highly enriched uranium for at least three nuclear weapons. On top of this the 1 per cent enriched tails would still weigh about 2 440 kg (98 per cent of the original feed) and these might be surreptitiously returned to the point of diversion and either mixed with large amounts of normal feed material or fed into the cascade at an appropriate stage. If this were done successfully, the diversion could be quite effectively concealed.

This calculation calls into question another critical number used to guide safeguards applications, the so-called conversion time [36b]. This is the estimated time it would take to convert a given form of nuclear material to the metallic components of a nuclear explosive device. It is also closely related to the detection time used to establish the timeliness criterion implied in the safeguards objectives. This criterion is used to establish inspection and inventory frequencies, and containment and surveillance measures [36c]. The estimate of this time is intended to be conservative, so it assumes "that all necessary conversion and manufacturing facilities exist, that processes have been tested, and that non-nuclear components of the device have been manufactured, assembled and tested" [36b].

The conversion time necessary for low-enriched uranium is assumed to be of the order of one year [36d]. In light of the capabilities computed above for a centrifuge plant and of the prospects for even more rapid and compact techniques for achieving high enrichments, it would seem prudent to re-evaluate this criterion as soon as possible.

---

[1] For comparison, note that a typical shipping cask for enriched UF_6 can hold 10 000 kg, or almost 7 000 kg of uranium [37a].

[2] The calculation on which these estimates are based is done in chapter 5 following table 5.1.

The success of the above diversion scenario would depend on an ability to accomplish the diversion and return the tails between IAEA inspections. An obvious way to prevent this is to have inspectors in residence continuously at all enrichment facilities. This is one of the recommendations made in chapter 4.

Another possible means of detecting the above diversion might be to employ the recently proposed minor isotope safeguards technique (MIST) [38]. This technique relies on the presence in natural uranium of very small quantities of another isotope, ^{234}U (see footnote 1, p. 3) and/or the presence in irradiated reactor fuel of the isotope ^{236}U. In an enrichment stage the same physical processes which operate to change the proportions of ^{235}U and ^{238}U also cause changes in the relative amounts of ^{234}U and ^{236}U. However, since the stage separation factor for each isotope is different, the degree of enrichment or depletion per stage is different for each species. This leads to a complex distribution of relative isotopic compositions over the whole cascade and to isotopic compositions of product and tails which can reveal a great deal about the detailed operation of the cascade.

For example, in the above diversion example, if only ^{235}U and ^{238}U were present in the uranium being processed, there would be no way of learning from the external product and tails enrichments that an extra feed stream was being added at some higher level of enrichment than the natural feed. But if the ^{234}U and ^{236}U compositions of the tails and product were being monitored as well, such an added feed stream could in principle be detected [38a].

Whether it could be detected in practice is another matter. If the size of the diversion were very small compared to the throughput of the plant, then the isotope sampling techniques would have to be very precise to see the changes created by the modification of the feed patterns. So far the few experimental tests of MIST which have been conducted have suggested that much more sophisticated mass spectrographic techniques will have to be developed before even relatively large modifications can be reliably detected [38b].

Material accounting

The methods currently used to detect diversion involve " . . . materials accountancy as a safeguards measure of fundamental importance, with containment and surveillance as important complementary measures" [6c].

The material accountancy function at an enrichment facility would be carried out as follows [36e].

1. *Dividing nuclear material operations into material balance areas (MBAs).* These are areas for which the quantity of nuclear material transferred in and out can be monitored and the total amount present (the physical inventory) can be determined. At an enrichment facility the

MBAs would most likely be the cascade itself and the rooms in which feed, product and tails material are stored and handled. In some kinds of enrichment plant this could all be done in a single room.

2. *Maintaining records describing the quantities of nuclear material held within each MBA.* This implies the taking of periodic inventories which account for the total weight of uranium present along with the isotopic content of each batch. A batch in this context would refer to a single cylinder of UF_6 [36f]. The cylinders can be weighed very accurately on scales which are capable of determining the weight of uranium in a $10-14$ ton cylinder [37a] to within ± 0.5 kg [37b]. The isotopic content is measured by mass spectroscopic techniques, which have inaccuracies of the order of 0.1 per cent, mainly due to uncertainties in the isotopic standards used in the mass spectrographs [37c]. It may also be possible to use instruments which measure the isotopic composition of the material inside cylinders by monitoring the gamma ray emissions from the two isotopes [33a]. This latter technique shows promise of allowing more rapid spot checks, and may be used by IAEA inspectors for on-site sampling, although it does not seem to be capable of as high accuracies as mass spectroscopy. In any event it can be said with considerable assurance that the amount of ^{235}U present in cylinders of UF_6 can be determined with accuracies well within the IAEA standards of 0.2 per cent (see point 5 below).

3. *Measuring and recording all transactions involving the transfer of nuclear material . . . from one MBA to another or changes in the amount of nuclear material present due to nuclear production or nuclear loss.* These measurements are taken at "key measurement points" (KMPs) which are locations where nuclear material appears in such a form that it may be measured to determine material flow or inventory [36g]. The number and locations of KMPs is an important and potentially controversial issue at enrichment facilities. This point is discussed in detail below.

4. *Periodically determining the quantities of nuclear material present within each MBA through the taking of physical inventory* (see point 1 above).

5. *Closing the material balance over the time period spanned by two successive physical inventories and computing the material-unaccounted-for (MUF) for that period.* Between physical inventories the record of inputs and outputs from the various MBAs is kept in a book inventory. At the next physical inventory the book value is compared with the actual value to determine the amount of MUF.

The closing of a material balance involves comparing the current inventory with the previous one and verifying that any difference is accounted for by known inputs and outputs. This process must be both accurate and fast to satisfy the criteria for timely detection of significant diversions. In point 2 above it was noted that weighting and sampling techniques on UF_6 cylinders are quite accurate. However, in practice the gathering, processing and reporting of this information to the IAEA is more time-consuming than it ought to be under the timeliness standard. According to one assessment, a period of at least one month elapses between the time a material balance is closed and the IAEA can begin to officially receive the processed accountability data. As a result: "By the time the IAEA has independently analyzed samples, it may be verifying operator data that were measured two or three months earlier" [39]. In view of the possibility of the 18–day diversion example described above, this situation is clearly unsatisfactory.

6. *Providing for a measurement control programme to determine accuracy of measurements and calibrations and correctness of recorded source and batch data.* Highly accurate measuring instruments are crucial to the success of a material accountancy programme, especially as the amounts of material processed between inventories grow. Large enrichment facilities process material at a greater rate than any other type of facility in the nuclear fuel cycle. Therefore, the precision and frequency of flow and inventory measurement take on a special significance in the case of enrichment plants. In this connection it is interesting to note the following rather divergent assessments of the state of the measurement art:

"The measurement techniques for UF_6 at the material balance boundaries of gaseous diffusion, nozzle and centrifuge plants are highly accurate, thereby enhancing the ability of materials accountancy to detect the possible diversion of a significant quantity of nuclear material which is reflected in MUF, with the degree of detection sensitivity dependent, *inter alia*, on the plant through-put and any changes in hold-up". [13b]

"Consider materials accountability. The earliest safeguards studies led to the conclusion that materials accounting measures . . . were subject to an unavoidable limit in the accuracy of materials balances [which] . . . when applied to facilities to large inventory or through-put, could mask diversions of sizeable proportions. Today, despite giant strides in measurement, analytical and statistical techniques, this qualitative conclusion remains unchanged". [40]

These two statements seem to be making very different assessments of the capabilities of materials accountancy measures, but the difference is more in tone than in substance. In spite of its obvious desire to put the best face possible on the existing techniques, even the International Nuclear Fuel Cycle Evaluation (INFCE) (first) statement cannot avoid mentioning that the sensitivity of detection depends on the sizes of plant throughput and/or inventory. This means that large enrichment plants will have large absolute uncertainties in these important measurements, and that small diversions from large facilities will always be difficult to detect by pure accountancy techniques.

7. *Testing the computed MUF against its limits of error for indications of undetected loss.* This is the point at which an MUF greater than the expected error limits could first raise suspicions that a diversion had taken place since the last physical inventory. As noted above, this discovery is likely to be made several months after the diversion, if at all.

8. *Analyzing the accounting data to determine the cause and magnitude of mistakes in recording, unmeasured losses, accidental losses and unmeasured inventory (hold-up).* This last item, the hold-up, represents another problem with special relevance to uranium enrichment facilities. As an example, consider a large (10 000 t SWU/yr) gaseous diffusion cascade used for making 3 per cent enriched product. Use of tables 6.3 (p. 188) and 5.1 (p. 113) shows that the inventory of uranium in the cascade itself is 1 200 tonnes, and the throughput is 12 700 tonnes per year.

The 'expected operator measurement uncertainty associated with closing a material balance' is defined by the IAEA as 0.2 per cent of the larger of inventory or throughput [36h]. (The origin and significance of this standard will be discussed below.) If it is assumed that the balance is closed twice per year at the above gaseous diffusion plant, the expected error would be 0.002×6 350 or 12.7 tonnes. This is five times as large as the significant quantity of 2.5 tonnes which the system must be able to detect (see p. 45). A straightforward solution to this problem is to check material balances more often, so that smaller quantities are involved. For example, checking once per month would reduce the uncertainty in MUF to 2.1 tonnes, just within the acceptable limit. However, quite aside from the inconvenience and expense of taking inventory this often, it should be noted that the size of the hold-up in the cascade, and therefore its measurement error, does not depend on the time interval between inventories. This puts a limit on the accuracy which can be achieved in any material balance, no matter how short the time between balances. For the facility under consideration this lower limit is 2.4 tonnes, just barely within the acceptable range. A chemical enrichment facility using solvent extraction would have a hold-up more than 15 times as large as a comparable gaseous diffusion plant (see table 6.3). For such a facility this problem becomes much more serious.

This discussion so far has assumed that it is possible to physically measure the amount of uranium inside an enormous cascade to an accuracy of 0.2 per cent without any observations on individual cascade elements or any data on internal cascade flows, power levels, or operating parameters. These restrictions follow from the customary practice of refusing IAEA inspectors access to the cascade area, a restriction rationalized by the desire to protect industrial secrets and recognized as legitimate in IAEA statutes [6d]. In practice this means that any independent verification of the cascade inventory by the IAEA will have to be based entirely on data taken outside the cascade. There is little evidence that this can be done with the precision necessary to satisfy the IAEA criterion of 0.2 per cent relative error.

There are a few proposals for improving cascade inventory measurements. One which has been advanced for the Brazilian jet nozzle facility is to add extra key measuring points at various locations in the plant to improve the accuracy of inventory measurements [41]; but it is important to notice that none of these extra KMPs are inside the cascade itself. This means that the cascade hold-up measurement must still be a dominant factor in determining the final MUF error estimate. This may be the reason why, even after the proposed additions, the measurements are found in a first, rough calculation to be reliable to only 0.2–0.3 per cent of feed flow in a biannual inventory [41a]. It should also be noted that the proposed Brazilian facility will have a capacity of only 200 t SWU/yr.

Another proposal for a non-intrusive inventory measurement technique relies on the use of minor isotopes similar to the MIST technique described above [42]. The method is derived from the theoretical observation that for any cascade there is a direct proportionality between the inventory and the equilibrium time. A cascade which has come to equilibrium is given a sample of feed which has been 'spiked' with one or both of the minor uranium isotopes. Then the product and tails streams are monitored for a period of time long enough to observe that most of the added isotope has passed through the system. The time taken for this to occur is then compared with a similar time computed for a model cascade with a known inventory. This comparison allows a computation of the unknown inventory of the actual cascade.

The method is theoretically quite elegant, and an initial experimental test on the Oak Ridge gaseous diffusion cascade gave an experimental error of about 3 per cent [42a]. Although the experimenters considered their result to be a very good one, it should be emphasized that this experimental error is 15 times as large as the standard set by the IAEA. It is not possible to judge how likely it is that the method could be improved to meet this standard at an acceptable cost.

A solution to the cascade inventory problem would be to allow access to the cascade area by inspectors, or to place inventory measuring devices inside the cascade itself. The latter could be done by installing monitors for pressure, temperature and isotopic composition of the process gas at each stage of the cascade and using data from these instruments to calculate the inventory. Such methods are already used at the US gaseous diffusion plants to aid the plant operators in keeping material balances [43], but the results are kept secret [42b] and no data on accuracies are available. However, even if the operator were able to approach the 0.2 per cent level of accuracy using these classified data, the IAEA would still have to find some alternative means of independently verifying such inventories. According to one US official, "it remains to be seen to what extent the IAEA can in practice verify independently the operators' measurements" [43].

This analysis leads to the conclusion that there is no substantial empirical foundation for the IAEA measurement standard of 0.2 per cent of throughput or inventory. It is not clear how this number was arrived at, but it is clear that substantial technological and political barriers must be overcome before it can be considered realistic.

The analysis also leads to the somewhat paradoxical conclusion that processes like gaseous diffusion and chemical exchange, which, because of their huge inventories and long equilibrium times, seem the *most* secure against conversion to highly enriched uranium, are the *least* secure against the undetected diversion of significant amounts of low-enriched material. This suggests that some very optimistic statements about the proliferation resistance of chemical-exchange processes should be read with caution [44a].

An important feature of materials accountancy as applied by the IAEA is that it depends heavily on the accounting system kept by the state [6d]. This implies that permanent IAEA staff are not present to observe all measurements and record all data. The function of the IAEA inspectors is to make periodic routine visits to the facility during which they audit the inventory accounts, make spot checks of inventory and materials assays and verify the calibration and accuracy of instruments [6e]. They are also allowed in principle to make unannounced visits, but in practice this presents certain problems both for the inspectors and the facility operators [30c]. It is impossible to determine on the basis of public information how often such unannounced visits are made.

Containment and surveillance

From this brief review it is clear that material accountancy methods can serve a useful function in monitoring nuclear materials and in creating obstacles to easy diversion. But it is also evident that these methods have weaknesses, and that some of these weaknesses are particularly acute in the case of uranium enrichment facilities. Therefore, it seems essential that material accounting measures be supplemented by substantial efforts at containment and surveillance.

As was stated above, these are seen by the IAEA as "important complementary measures". Perhaps one can get some idea of how truly 'complementary' they are by noting that in the IAEA Safeguards Glossary 22 pages are devoted to material accountancy measures and their implementation, while only 3 discuss containment and surveillance. One IAEA official has emphasized that some states reject effective containment and/or surveillance measures, and this puts a disproportionate share of the verification burden on accountancy [45].

Containment involves the use of "physical barriers; e.g., walls, transport flasks, containers, vessels, etc., which in some way physically restrict or control the movement of or access to nuclear material and to IAEA surveillance devices" [36i]. Aside from the last-mentioned use, the isolation of surveillance devices, it is difficult to see how containment could play much of a role in safeguarding an enrichment facility. The flow of materials is more or less continuous and, just as in any other industrial operation, the facility managers will want to keep product inventories as low as practical and have convenient access to these inventories. It may be possible to work out some kind of containment schedule based on known production and delivery schedules, but one suspects that the facility operators would strongly resist such infringements on their flexibility.

One possible use of containment measures would be the restriction of access to the cascade area to only a small number of monitored entrances. Other doors, such as fire and safety exits, could be sealed in such a way that they could be used if needed, but that any use of them would necessarily be detected. This could make it more difficult to modify a cascade without detection.

Surveillance involves "the collection of information through devices and/or inspector observation in order to detect undeclared movements of nuclear material, tampering with containment, falsification of information related to locations and quantities of nuclear material, and tampering with IAEA safeguard devices" [36j]. It is easy to think of many ways of using surveillance techniques to increase substantially the difficulty of either diverting nuclear material or modifying the operation of a facility clandestinely. Sealed television cameras, optical, acoustic and seismic sensors, resident IAEA inspectors or observers, continuous metering of process flows and power consumption, and so on, could all be combined to make it virtually impossible for any unauthorized activity to escape detection.

However, this generous menu of technological possibilities must be tempered by economic and political constraints. The more elaborate and foolproof the surveillance system, the higher the cost, and the question of who shall pay the costs of implementing safeguards has been a contentious issue for many years [30d]. There is also the question of the testing and maintenance of surveillance equipment. If this required frequent access by IAEA inspectors or maintenance people, then the operators could claim that the system was interfering with the efficient operation of the facility, something safeguards are explicitly forbidden to do.

But above all there are the political issues of equal treatment and industrial secrecy. Given the highly asymmetrical character of the NPT itself and the manifest inequality of treatment of states with and without nuclear weapons, it is not surprising that states resent and resist the highly intrusive and implicitly insulting application of surveillance measures.

Alongside this is the desire to prevent the dissemination of industrial secrets. This is the reason most often cited for forbidding access by IAEA inspectors to the cascade area of enrichment facilities [15a]. This principle of exemption of certain areas from inspection has even been incorporated into the formal guidelines for safeguards agreements [6c]. It would seem to be implicit in this exemption that most kinds of surveillance techniques would also be excluded from the cascade area. This would rule out the sorts of monitoring and recording devices which might be used to more accurately determine cascade hold-up, as well as surveillance devices which could detect changes in piping or operation within the cascade area. Detection of such activities can then only be made through materials accountancy methods, supplemented by such non-intrusive surveillance measures as are permitted by the state.

This situation can be improved somewhat if the IAEA has sufficient information on the design, layout, and operating modes of the plant to be able to make inferences from purely external data about what may be going on inside the restricted area. Provision for supplying such data as early as possible to the IAEA is called for in all safeguards agreements, but the data are to be restricted to rather general information, and only that design information which is relevant to the application of safeguards [6f]. It is stated that the design data should be "in sufficient detail to facilitate verification", but the same paragraph also contains the important exemption referred to above. Whether or not IAEA inspectors would be able to infer from materials accountancy and external surveillance measures that cascade modifications or batch recycling operations had been carried on inside the restricted area remains, therefore, an open question.

Summary

An evaluation of the usefulness of safeguards at enrichment facilities must begin with the understanding that INFCIRC/66/Rev 2 and INFCIRC/153 are political documents in which the capabilities of technology have been subordinated to the process of political negotiation. This is as it should be, and as much as one might wish for a political atmosphere more conducive to the constructive and efficient implementation of safeguards, such political changes cannot be brought about by technological means, no matter how creative.

A second point which needs to be emphasized is that, for all of their weaknesses, the safeguards measures now used by the IAEA do significantly complicate the process of clandestine diversion of nuclear materials. They also increase the risk of detection to a level where it is reasonable to conclude that only a clever and determined, or possibly desperate, group or state would attempt such a diversion.

It can be concluded that if all enrichment facilities were placed under IAEA safeguards the risks of this route to nuclear proliferation would be substantially reduced. But it is enough to state this requirement to understand why the safeguards approach is so limited. Most of the world's enrichment capacity is not now and seems highly unlikely to come under IAEA safeguards in the foreseeable future. Safeguards can only work where they are applied, and they are applied sparingly.

Most assessments of safeguards end with the conclusion that they are a necessary, but insufficient, component of any non-proliferation regime. Our assessment is no different in this respect. What we have added here is a better appreciation of the special difficulties involved in applying safeguards to uranium enrichment facilities. These make it clear that safeguards will never be any better than the political and institutional framework in which they are employed. It is on this framework that attention will now be focused.

III. Institutional control mechanisms

Institutional measures as they will be defined in this section are non-technical in nature and involve various political, economic or diplomatic strategies for controlling access to sensitive materials, facilities or technology. Obviously the safeguards regime as administered by the IAEA deserves to be called an institutional measure, but it has been separated for special treatment because of its primarily technical nature. The measures to be discussed below generally lack this technical component.

The following analysis of institutional measures is organized according to the degree of international co-operation involved in their implementation. The spectrum ranges from the purely unilateral through multinational arrangements of varying sizes to fully international efforts. An attempt is made to cover all of the measures which have been tried or seriously proposed for the future. In this analysis it will be assumed that the reader has some familiarity with the history of non-proliferation efforts and the state of the world enrichment industry. For those readers who wish to review these areas, a more detailed historical survey is given in Part Three along with a country-by-country status report on enrichment activities and non-proliferation policies. The material in Part Three provides much of the factual basis on which the following analysis depends.

Unilateral measures

During the 1970s it began to be more widely recognized that safeguarding the facilities which provide direct access to weapon-usable material is not sufficient to prevent proliferation. This led logically to the idea that the spread of the facilities themselves to other countries should be limited. One of the ways of enforcing such limitations is by means of unilateral measures. Actually, with respect to enrichment facilities, this has been almost without exception the policy of the nuclear weapon states.

The character of unilateral measures can vary widely: countries can be discouraged from building their own facilities by offering them enriched uranium on attractive terms; technology can simply be denied to other countries; economic sanctions can be employed against a country that threatens to build its own enrichment facility; or an enrichment facility can be attacked and destroyed by a country which feels threatened by it. This latter tactic, although hardly in the category of 'institutional' measures, was added to the list of active options by the June 1981 Israeli attack on Iraq's research reactor. There is also evidence that pre-emptive measures have been contemplated by other nations in the past. It has been alleged that the USSR approached the USA with suggestions for a collaboration or at least a tacit approval of a Soviet strike against Chinese nuclear installations [46]. And as long ago as 1948 it was prominently suggested that the United States should attack the Soviet Union before the latter could develop nuclear weapons [47]. It is reliably reported that the

Pakistani centrifuge facility, now under construction, is heavily defended against air attack [48, 49].

We will not analyse the virtues and defects of this mechanism for proliferation control in detail here. Suffice to say that this form of violent, unilateral action is hardly less threatening to world peace and security than the proliferation it seeks to prevent. If there is to be any genuine progress towards non-proliferation, it seems clear that it will be made by other means.

The historical review in chapter 7 shows that several unilateral measures have been tried in the past. After World War II the USA first tried to keep enrichment technology secret and to prevent transfer of such facilities to other countries. After 1953 this policy was continued, but supplemented by the offer of enrichment services to other countries, aimed at discouraging independent enrichment developments. With only one exception — the aid given to China in the 1950s — the USSR pursued a similar policy. The Nuclear Suppliers Group (the so-called London Club) agreed in 1976 to exercise restraint in the transfer of sensitive facilities. In a number of cases this would mean denying technology to other countries. The same restraint is implied in the 1976/77 US, French and West German embargo on future exports of reprocessing plants.

The US International Security Assistance Act of 1978 (Public Law 95–92) goes even further and provides for economic sanctions to prevent other countries from constructing their own enrichment plants. Concerning enrichment the Act states that the USA is to cut off funds for economic and military assistance to any country that delivers nuclear enrichment equipment, materials or technology to any other country or receives such items from any other country unless two conditions are met before delivery: (a) such items must upon delivery be put under multilateral auspices and management when available; and (b) the recipient country should accept full-scope safeguards [44a]. In accordance with this law, US military aid to Pakistan was terminated in April 1979 when it became evident that Pakistan was building a clandestine enrichment plant with centrifuge technology illegally obtained from the Netherlands. However, after the Soviet intervention in Afghanistan in 1979, President Carter proposed to resume the military assistance based on consideration of the vital interests of the USA, a possibility kept open in the law. In October 1981 the US Senate approved the resumption of military aid to Pakistan, subject only to periodic reassurances by the US President that Pakistan is not seeking to build nuclear weapons.

The London Club of nuclear supplier states is included in this analysis of unilateral measures because of its informal and one-sided nature. The motivations for the formation of the London Club are discussed in chapter 7, so here it will be characterized simply as an effort by a group of nations which were in a position to supply sensitive nuclear materials or technology to exercise restraint in doing so to non-nuclear weapon states. This agreement was the result of a series of secret meetings among the supplier states in 1976–77.

The major results of the negotiations of this group of supplier states were a list of sensitive nuclear materials, equipment, facilities and technology and a set of guidelines for the export of such items to non-nuclear weapon states. The list of sensitive items has been called a 'trigger list', since export of any item on this list was intended to trigger the application of IAEA safeguards. Except for the addition of heavy-water production plants, and for the explicit listing of what are to be considered 'major critical components' of various enrichment processes, this list is identical to the so-called 'Zangger trigger list', already in use since 1974 by the IAEA in interpreting the NPT safeguards provisions [50a, 51a].

The main London Club guideline states that the spread of sensitive facilities should be limited as follows:

Suppliers should exercise restraint in the transfer of sensitive facilities, technology and weapons-usable materials. If enrichment or reprocessing facilities, equipment or technology are to be transferred, suppliers should encourage recipients to accept, as an alternative to national plants, supplier involvement and/or other appropriate multinational participation in resulting facilities. Suppliers should also promote international (including IAEA) activities concerned with multinational regional fuel cycle centres. [52a]

One new element in this guideline which deserves emphasis is the suggestion that IAEA safeguards are not only to be applied on transferred nuclear material, equipment, and facilities but that these safeguards should also apply to sensitive facilities for which only the technology has been transferred. Technology in this context apparently means technical data in physical form considered to be important for these facilities and not available in open sources [50b]. According to the guidelines IAEA safeguards should also apply to sensitive facilities of the same type (i.e., "based on the same or similar physical or chemical processes") not necessarily transferred, but constructed during an agreed period (generally 20 years) in the recipient country [52b].

The guidelines also contain a provision requiring any country receiving sensitive items to demand the same assurances as required by the original supplier from any other country before any retransfer of sensitive items can occur. In addition, the supplier's consent should be required for the retransfer of sensitive equipment, facilities or technology, and for the transfer of items derived therefrom. This also holds for the retransfer of heavy water and of weapon-usable material.

A special provision was included on the transfer of enrichment facilities, or technology therefor, stating that "the recipient nation should agree that neither the transferred facility, nor any facility based on such technology, will be designed or operated for the production of greater than 20 per cent enriched uranium without the consent of the supplier nation, of which the IAEA should be advised" [76a].

The export guidelines of the London Club were the result of a compromise of strongly conflicting interests and motives. Some of the more obvious were: (a) a desire to prevent the proliferation of nuclear weapons and to reach agreement among the major nuclear supplier countries on a stricter non-proliferation policy; (b) a desire to stimulate trade in nuclear equipment and technology for civilian use; (c) a US desire to prevent the bargaining of non-proliferation conditions as part of the competition for nuclear-export contracts; (d) a French and West German desire not to change existing export contracts which already included the transfer of sensitive technology; and (e) a desire by all parties to maintain and enhance their economic and technological positions in the nuclear field.

The best compromise obtainable was an agreement by the supplier countries to exercise restraint in the transfer of sensitive facilities and technology, and of weapon-usable material. This carries the clear implication that the supplying countries need not entirely refrain from such exports. The result was that France in 1980 still delivered highly enriched uranium for the new Iraqi research reactor which was subsequently destroyed by Israel, and that in the same year FR Germany and Switzerland signed a contract with Argentina to supply a heavy-water reactor and a heavy-water production plant. Another weakness of the export guidelines is the absence of any requirement for full-scope safeguards for the recipient countries, a provision which the United States and the Soviet Union tried very hard to obtain. The specific safeguards conditions agreed to by the London Club were a generally weaker substitute for the more rigorous full-scope safeguard condition. The latter would have made the former redundant. Only the requirements for restraint on export of enrichment and other sensitive technology, and for prior consent, went further than the full-scope safeguard condition, in the sense that they would limit possible physical access to weapon-usable material.

The London Club's agreement was much criticized by a number of Third World countries, who accused the supplier countries of forming a kind of suppliers cartel. In addition, it was alleged that the export guidelines when applied to NPT parties were inconsistent with Article IV of the NPT, which recognized the right of all NPT countries to participate in "the fullest possible exchange of equipment, materials and scientific and technological information for the peaceful uses of nuclear energy".

As a one-sided measure by a group of nuclear supplier countries, the London Club guidelines are much less universally accepted than the NPT. Actually the guidelines might even encourage second-echelon nuclear countries to pursue independent nuclear development more vigorously. Moreover, the London Club guidelines were focused only on measures to prevent horizontal proliferation, neglecting any possible connection with the vertical proliferation problem. The NPT at least addresses this latter problem, even though a number of its members only pay lip service to this provision. In conclusion, while it can be argued that the London Club measures can delay the spread of sensitive nuclear facilities and the proliferation of nuclear weapons through the use of these facilities, these measures in themselves certainly cannot stop that process. In fact such discriminatory measures can backfire. If nuclear energy turns out to be desirable to many countries, and if no arrangements are made for assured fuel supply at fair prices, then attempts to restrict exports by cartel-like mechanisms are most likely to stimulate further indigenous developments of sensitive technology.

In recognition of these dangers the US Nuclear Non-Proliferation Act (NNPA) of 1978 sought to combine the carrot of assured fuel supply with the stick of demanding full-scope IAEA safeguards in the recipient non-nuclear weapon countries (see chapter 7). In order to give this policy credibility the USA proposed an expansion of its enrichment capability.

The expansion of US enrichment capacity was intended to have two major effects. First, a guarantee of the supply of enriched uranium would eliminate the need for plutonium recycling and reprocessing of spent fuel. Second, the prospects for implementation of US non-proliferation objectives would increase if other countries remained dependent on US enrichment services. The latter could, of course, only be effective if no alternative suppliers of enrichment services were willing to be more lenient on non-proliferation conditions or if enrichment services were in short supply.

France, FR Germany and Japan strongly protested against the NNPA, even more strongly than they had against the 1977 anti-plutonium decision (see chapter 7). Under the NNPA these countries now had to obtain consent from the USA for reprocessing spent fuel which originated from US nuclear supplies. When in 1978 Euratom initially refused to renegotiate the existing nuclear agreements, the USA temporarily cut off enriched-uranium supplies as required by the NNPA. Because European enrichment capabilities were at that time still not large enough to fill the resulting supply gap, the operation of a number of European nuclear power plants was seriously threatened by this embargo. The supplies were resumed after a compromise agreement was reached to await the results of the International Nuclear Fuel Cycle Evaluation (INFCE) (pp. 78–79). It should be noted that the INFCE was completed in February 1980, but that no effort has been made by the USA to revive the embargo.

The unilateral measures of the NNPA also met strong protests from many other countries. Serious conflicts arose between India and the United States when the continuation of enriched-uranium supplies to India came under attack in the US Congress in 1980 because of India's refusal to accept full-scope safeguards. It required strong lobbying by the Carter Administration to convince Congress to allow the delivery of enriched uranium to India [53, 54, 55, 56].

One more unilateral measure which deserves analysis is self-denial. By choosing not to develop or employ a type of technology a nation can not only help to retard the spread of that technology, but can also serve as an example of the virtues and benefits of getting along without it. The Carter

Administration's 1977 anti-plutonium decision, in which the USA deferred indefinitely commercial fuel reprocessing, might be considered to be such an attempt (see chapter 7). Apparently, the United States was aware that it could not reasonably demand of other countries that they abandon their efforts towards a plutonium economy while at the same time continuing its own programme. This might be one thing that had been learned from the NPT experience, where the discriminatory character of the Treaty has been one of the major obstacles to its universal implementation.

However one might evaluate the success of the anti-plutonium policy, it is clear that abstaining from uranium enrichment is simply not compatible with exploiting nuclear energy on the basis of light water reactors. A rejection of enrichment would only be compatible with the use of natural uranium reactors, such as those moderated by heavy water or graphite, and there is little prospect of a widespread change-over to these alternatives in the foreseeable future. Some kind of world enrichment industry will be necessary for many years to come, and given the heavy investment already made by a number of countries and the existence of many long-term contracts for enrichment services, it is not realistic to expect any such drastic unilateral actions by any of the current suppliers of these services.

However, it is much more reasonable to suggest that nations refrain unilaterally from developing and using certain enrichment processes which are particularly proliferation-prone.

The molecular laser isotope separation process is a case in point [57]. Given the still early stage of development of this process and the considerable effort and expense which will be necessary to demonstrate its feasibility, a decision by one or more of the countries currently developing the process to drop it could have an important impact, both practical and psychological, on future non-proliferation efforts. Such a decision might help to undermine the fatalistic and self-fulfilling notion that technological progress will inevitably undermine all political efforts at control.

How effective are unilateral measures in preventing the spread of sensitive facilities? History shows that in the short run they may certainly have some effect but that in the long run they cannot prevent the independent development of these facilities and the associated nuclear weapon capabilities. The UK and France provide early examples, both having developed their own gaseous diffusion plants (see figure 3.1), the latter in the face of strong opposition by the USA. The Netherlands and FR Germany several years later were able to develop their own centrifuge enrichment technology. FR Germany has also developed the jet nozzle method and South Africa its Helikon enrichment technique. Of all the countries which now possess enrichment plants only China, and to some extent South Africa and Pakistan, depended on transfer of the technology from foreign countries (in Pakistan's case an important part of this transfer was, of course, involuntary). Most of the spread of enrichment capability has been the result of indigenous developments. Thus, it is fair to conclude

Figure 3.1. An aerial view of the uranium enrichment plant of British Nuclear Fuels Limited at the Capenhurst Works, near Chester

The centrifuge enrichment plant, which was opened in 1977, is on the right-hand side of the photograph. The long building running across the centre is the diffusion enrichment plant which began operating in the 1950s to enrich uranium for military purposes.

Source: British Nuclear Fuels Ltd.

that the best that can be expected from unilateral non-proliferation measures is that they delay the dissemination of technology. This in itself might be a positive effect if the resulting delay were used constructively to develop more effective international policies, but in the absence of such developments the delay can have no ultimate value. And unilateral measures can be politically damaging, leading to tensions both among supplier nations and between suppliers and consumers. The experiences of the London Club and the US Nuclear Non-Proliferation Act give ample evidence of these dangers.

Multinational facilities

Basic idea

A multinational enrichment facility is one in which ownership, control and/or operation are shared among a number of nations. For several reasons such an arrangement could, if appropriately arranged, reduce the risk of clandestine production of highly enriched uranium and of diversion of nuclear material from the enrichment plant.

First, the participants in a multinational facility can watch each other, making the diversion of nuclear material politically more risky [51b, 58a, 59a, 60a]. This could reduce the risk involved in denying safeguards inspectors access to the facility or to its sensitive areas [58b]. The possibility of seizure of the plant by the host country would always be present, but because of the resulting confrontation between this country and the other participants, a considerable political barrier inhibits such an action.

Second, the use of multinational facilities instead of a multiplicity of national facilities would reduce the number of plants to be placed under safeguards, increasing the feasibility of continuous inspection [15b, 58b, 59b]. If several countries from the same region participated in a regional multinational facility, this could also help to reduce suspicions in these countries about their participating neighbour's nuclear weapon intentions [58b].

The multinationalization concept presumes the existence of several mutually independent multinational enterprises. When not restricted to supplying their enrichment services to the participating states (in contrast, for example, to a strictly regional arrangement) these enterprises will compete with each other for contracts with non-participating countries [13c, 51c]. It has been argued that a competitive world market to which consumer countries have free access would enhance the security of enriched-uranium supply to these countries [13d, 13e]. However, such a situation could just as easily lead to a competition among suppliers regarding the scope and strictness of safeguards conditions on enriched-uranium supplies.

Incentives for multinational arrangements need not, and in practice mostly do not, originate from non-proliferation objectives [15b, 61a]. Other objectives present in a number of European joint enterprises were: technological and commercial opportunity, reliability (or independence) of nuclear fuel supply, financial and economic risk-sharing, and international organization interest (e.g., joint activities within the framework of Euratom or the Nuclear Energy Agency of the Organization for Economic Cooperation and Development, NEA/OECD [61b] (see also chapter 7).

The structure of multinational arrangements

No single concept exists for multinational management of the sensitive parts of the nuclear fuel cycle. For reprocessing, the concept of a regional fuel centre is usually considered to involve only participation of countries from the same region [59] in order to reduce transport of sensitive nuclear materials. However, this certainly is not the case for multinational uranium enrichment arrangements.

Because of differences in objectives, multinational enterprises may have different institutional forms, which in turn may have rather different implications for non-proliferation efforts. Thus, it is possible to imagine an arrangement with only financial/commercial participation in an enrichment plant, while the plant itself is located in a single country and operated exclusively by personnel from that country without any sharing of technology. A different institutional form might involve much more thoroughly multinational management and plant operation with substantial sharing of technology.

In establishing a multinational enrichment enterprise many questions must be resolved, including, *inter alia*, the conditions for participation, number and location of plants, operation with national or multinational personnel, ownership, management and control of facilities, financing, conditions for supplying enrichment services, host country responsibilities and rewards, insurance and liability, health, safety and environmental protection, physical security, international safeguards and transfer of technology [51d, 62a].

Most of the fundamental issues must be settled at the beginning of the arrangement, partly by formulating principles to be observed when acting on these matters. Other issues which arise in the course of operation must be decided on an *ad hoc* basis. For these questions the structure of decision-making in the enterprise will be important, and this constitutes another crucial issue to be dealt with in creating a multinational enrichment arrangement.

From a non-proliferation point of view it has been strongly argued that a multinational consortium limited to corporations is inadequate; the political nature of non-proliferation issues demands participation by the governments of the states involved [13f, 13g, 59a]. It follows that the multinational organization should be established by a treaty which contains provisions for intergovernmental supervision, for example, by a political council, consisting of ministerial representatives [62b]. Such a political council should deal with and have responsibility for all matters relevant to proliferation questions, such as changes in consortium membership, policies on safeguards, classification of technology, research and development, plant location and so forth. This council would be analogous to the so-called 'Joint Committee' as it exists within Urenco. Adding such a bureaucracy to an industrial collaboration is unquestionably cumbersome, but the political stakes are high enough to make it necessary.

Conditions for non-proliferation effectiveness

Several features have been suggested for multinational arrangements as necessary or at least desirable for the effective pursuit of non-proliferation goals. Five of the most important are:

1. Multinational plants should be safeguarded by the IAEA in order to deter diversion of special nuclear material or clandestine production of highly enriched uranium [13h, 59c]. Governments of participating coun-

tries should be involved in projects by treaty, and multinational organizations should abide by common, for example, UN-established, codes of conduct.

2. This would rule out any competition among multinational organizations in matters relevant to proliferation issues [51e, 59d]. Withdrawal clauses like the one in the NPT are not advisable [51b, 62c], but some form of economic and/or political sanctions to be applied in case of violation of the treaty should be incorporated.

3. Multinational facilities should be substitutes for, rather than additions to, national enrichment facilities [58b]. Participation in the organization should restrict the right to enrich uranium nationally if not prohibit it all together [51f, 62d]. Research and development on enrichment technology in participating countries should preferably be performed only within the framework of the multinational arrangement. No support should be given to building national enrichment plants in other countries.

4. It is highly preferable that a multinational consortium should operate no more than a single enrichment facility located in one of the participant states, rather than several plants distributed over the participating countries. This would greatly facilitate the application of safeguards. In addition this measure would prevent a number of participating states from automatically obtaining their own national facility in case the multinational arrangement should break up.

5. Multinational facilities should preferably not produce highly enriched uranium [51g, 60b]. If highly enriched uranium is produced, such production should be limited to specific applications with no stock accumulation. This implies restricting the need for highly enriched uranium by phasing out or drastically reducing the number of reactors which use this dangerous material [62e, 63].

This list of desirable non-proliferation features is not particularly controversial. However, there are other features which might be added to the list on which opinions differ considerably. Among these issues the question of the sharing of sensitive technology among the participating countries is a particularly difficult one [51h, 59e, 62f]. It has been argued that for many countries access to such knowledge represents a potentially important industrial incentive to join a multinational enterprise [51h]. For example, refusals to share technology among participating countries party to the NPT could be, and have been, interpreted as being in conflict with Article IV of the Treaty, which encourages "the fullest possible exchange of equipment, materials and scientific and technological information for the peaceful use of nuclear energy". On the other hand, the sharing of sensitive technology would accelerate the dissemination of the technical capability to produce weapon-grade material throughout the world.

The dilemma is difficult to resolve. Whereas an unqualified policy of withholding sensitive technology could delay wider dissemination, it might at the same time make multinational fuel cycle arrangements more difficult, stimulating in the long run the development of national facilities. Thus, it has been argued that the critical choice is not so much between dissemination and non-dissemination, but whether, from the non-proliferation point of view, the gradual but relatively uncontrolled diffusion of enrichment (and reprocessing) technology leading to the establishment of more independent national plants is to be preferred over the deliberate, possibly swifter, but certainly more regulated transfer of technology under multinational auspices [51i]. In the latter case there is some assurance that the exploitation of the technology will be under multinational control. In any event, if a transfer of enrichment technology is agreed upon in a particular case, a simultaneous denial of the right of enrichment under national control seems imperative if proliferation is to be discouraged.

Evaluation of current multinational arrangements

With the above list of desirable non-proliferation features as a standard, it is now possible to evaluate each of the three existing multinational arrangements: Urenco, Eurodif and Euratom, with respect to the degree to which they incorporate these features (see table 3.1, p. 70). This allows an assessment of the current non-proliferation effectiveness of these arrangements and also provides a framework for suggesting improvements. Historical sketches of these three arrangements can be found in chapter 7.

Urenco. An important feature of this British–Dutch–West German collaboration is the use of more than one enrichment plant. Each of the participating countries actually has its own facility, although these are all located in the UK and the Netherlands. The West German pilot plant and first commercial plant were placed on Dutch territory (see figure 3.2) to calm political fears of a West German national enrichment capability. However, in 1978 the Joint Committee consented to the construction of one of the future extensions of the enrichment plant at Gronau on West German territory [64].

An important provision of the Almelo Treaty permits any of the parties to withdraw from the Treaty after its first 10 years, that is, any time after 1980. When this provision is combined with the decision to place enrichment plants on each country's territory, the result is the threat that a break-up of the co-operation will leave each of the participating countries with its own national enrichment facility. This is not just a hypothetical situation, since there has already been considerable tension among the partners concerning the nature of non-proliferation conditions to be applied to enrichment service contracts with other countries. A particularly difficult issue was the enrichment contract between Urenco and Brazil in the late 1970s. This caused considerable trouble within the troika because of the more stringent attitude to safeguards of the Netherlands, which required, *inter alia*, a guarantee of international supervision of the storage

Figure 3.2. The Urenco facility at Almelo, the Netherlands

One of the two small centrifuge halls is owned and operated by FR Germany. Almelo is located about 25 km from the Dutch-West German border, and less than 30 km from Gronau, the site of the planned West German centrifuge plant.

of separated plutonium in Brazil. This concern was caused by the West German–Brazilian nuclear contract, which included the construction of a reprocessing plant in Brazil.

In the absence of uniform, internationally agreed rules of conduct for enrichment services, future contracts may again lead the troika into difficulty. Another possible source of future trouble arises out of the differing attitudes towards nuclear energy among the three partners. In particular, the strong public anti-nuclear energy movement in the Netherlands could jeopardize the willingness of the Netherlands to continue its participation in the joint project.

With respect to the sharing of technology, the Almelo Treaty obligates each of the partners to accept participation of any other partners who wish to join new research and development projects on centrifuge technology [28a]. The Treaty restricts the right to build national gas-centrifuge enrichment facilities but does not absolutely prohibit such activities. More precisely, no facilities under national control based on the centrifuges used in the joint proposal are permitted, but national facilities based on different gas centrifuges which may be developed in the future are allowed if the other partners have refused to participate in the development of new technology. The participating countries are in no way restricted in their research and development on other isotope separation techniques, nor are they prohibited from constructing national enrichment plants based on other techniques.

The Treaty allows Urenco to produce highly enriched uranium with the restriction that it "shall not produce weapon-grade uranium for the manufacture of nuclear weapons or other nuclear explosive devices" [28b]. Until now the safeguarding of the enrichment plant has been carried out under Euratom auspices, and direct access to the plants has been denied to IAEA inspectors. The safeguards measures which have been instituted have been focused more on the Urenco plants in the Netherlands than on those in the UK. Negotiations for IAEA inspection of the commercial plants are now under way, but it is highly unlikely that they will result in permission for IAEA inspectors to gain access to the cascade area [15c] (see above discussion of safeguards, p. 49). As the negotiations have continued, a joint team of Euratom and IAEA inspectors has been carrying out the safeguards procedures since the end of 1979.

The Almelo Treaty obligates the partners to ensure that no techno-logical data, equipment, or nuclear material subject to regulation under the Treaty will be used by a non-nuclear weapon state to produce or otherwise acquire a nuclear weapon or other nuclear explosive device. However, Pakistan succeeded in clandestinely obtaining sensitive informa-tion from the Urenco plants in the Netherlands, information that is now being used by Pakistan to build its own centrifuge enrichment plant, possibly to produce weapon-grade uranium [56].

Urenco can be credited with one important contribution to non-proliferation. The Netherlands refused to consent to the West German proposal to sell centrifuge technology to Brazil as part of the above-mentioned nuclear package deal. However, as a substitute, FR Germany agreed instead to sell the less developed jet nozzle technology to Brazil, so the effectiveness of the Dutch refusal was undermined. However, if unanimous agreement exists, the Almelo Treaty permits the transfer of sensitive technology. Such a transfer is now under consideration as Urenco is negotiating with Australia for the construction of a centrifuge facility in that country.

Eurodif. Eurodif is a private commercial enterprise with no inter-governmental board supervising its conduct on non-proliferation matters. Of the five partners in Eurodif, two — France and Spain — are not parties to the NPT. The major, if not only, role of the partners is to provide the enterprise with capital and contracts for enrichment services and to participate in general production decisions [61c]. However, in 1980 a treaty regarding Eurodif was concluded between the governments of France, Belgium and Spain [65]. This was later joined by Italy but has not yet been endorsed by Iran. Although the agreement was concluded mainly for the benefit of tax concessions, it also contains some general provisions related to proliferation matters.

Under the Eurodif Treaty the parties have agreed to ensure that no sensitive information, equipment or special nuclear material obtained through the co-operation of Eurodif shall be used by a non-nuclear weapon state to produce or otherwise acquire a nuclear weapon or other nuclear explosive device [65a]. Suitable safeguards have to be accepted by recipient non-nuclear weapon states before these items may be transferred to these countries [65b]. These provisions are similar to those required by the Almelo Treaty for Urenco. Also similar to the Almelo Treaty is the provision in the Eurodif Treaty that no highly enriched uranium shall be produced for the manufacture of nuclear weapons or other nuclear explosive devices [65a].

In contrast to Urenco, Eurodif operates only one enrichment plant (see figure 2.2, p. 16). This is situated in France, which is already a nuclear weapon state. Technological and management control is completely in French hands, with the Commissariat à l'Energie Atomique (CEA; the French atomic energy commission) having full responsibility [61c]. No sensitive technology (such as diffusion barrier design) has been transferred to France's partners. However, some of the European partners have made substantial contributions to other parts of the Eurodif facility [61d, 66]. The arrangement does not preclude the participating countries from eventually undertaking independent national enrichment activities. The enrichment plant at Tricastin will be under Euratom safeguards, but so far no contract to this end has been concluded between Euratom and Eurodif.

Because of the slowdown of nuclear programmes in some of the participating countries and the consequent reduction of demand for enrichment services, the stability of the Eurodif partnership may be in doubt [61e]. After the 1979 Iranian revolution, when Iran virtually cut off its nuclear programme, the continuation of its participation in Eurodif came into question. While Iran still formally participates in Eurodif, financial disputes between Iran and Eurodif continue [67]. Italy is also a source of trouble for the consortium. Because of major cutbacks in its nuclear programme, Italy reduced its share in Eurodif from 25 per cent to 17.5 per cent and in mid-1980 was negotiating the sale of the remaining share to France (see chapter 8). The Eurodif arrangement contains no guarantees against a possible future disintegration of this multinational enrichment enterprise. However, even though the Eurodif Treaty is designed to expire on the same date as the Eurodif enterprise ends, the Treaty does state that its non-proliferation provisions should continue in force beyond this date.

Euratom. As a multilateral institution Euratom has two important connections with uranium enrichment. First, Euratom has been unsuccessful in establishing a unified multilateral European enrichment enterprise. French hopes for such a project were disappointed in the 1960s by the offer of cheap US enrichment services, and another attempt around 1970 also failed: instead of one joint European enterprise, two companies — Urenco and Eurodif — were established. Although European government leaders at the 1973 Summit Meeting in Copenhagen once again pledged to strive for "une capacité européenne d'enrichement de l'uranium recherchant un développement concentré et harmonieux des projets existants" (a European uranium enrichment capacity aiming towards a concentrated and harmonious development of the existing projects) [68], no such development seems likely in the foreseeable future. Second, the Euratom statutes give the Agency the right of option on all nuclear materials [69a], including the right of ownership of all special fissile material (which includes enriched uranium) and the exclusive right of concluding contracts related to nuclear materials coming from within or outside the European Community (EC) [69b][3]. This provision strongly resembles part of the original Baruch–Lilienthal proposals for international ownership and control of the nuclear fuel cycle. In practice, however, it turned out that the EC exercised the purchase option only on materials for relatively small common EC programmes and not for the much larger national programmes [70a]. After 1964 France no longer even recognized Euratom's jurisdiction and concluded agreements for fuel supplies without EC involvement [23c, 70b].

Thus, while on paper the Euratom arrangement looks like a substantial multinationalization of nuclear activities and especially of nuclear materials ownership, in practice it does not work out that way. Member states have in fact hardly been restricted in their initiatives and activities in the nuclear field. Technological developments and industrial

---

[3]   An exception is made for nuclear materials for defence purposes [69c], which is of special concern to France.

activities have been primarily national in character and as far as substantial co-operation among states is concerned, it is in most cases not co-ordinated by Euratom.

Summary and comparison

Table 3.1 summarizes the characteristics of the three multinational arrangements with respect to the non-proliferation features previously enumerated. It is clear from this summary that none of the three arrangements possesses more than a very small number of the desirable features. It also appears that the particular conditions which are fulfilled by each multinational arrangement are substantially different, and that a number of important conditions are not fulfilled by any of the arrangements. In addition, the Euratom Treaty is basically only a 'paper' arrangement. Except for safeguarding, it has in practice no actual control over the enrichment activities of its member states.

Urenco and Eurodif do not restrict their enrichment services to their own participants, but are competing for enriched-uranium contracts in the world market. This competition has in the past resulted in a relaxation of certain non-proliferation requirements. In practice this has resulted in requiring safeguarding on only the contracted enriched uranium (according to Article III of the NPT), but no full-scope safeguarding in countries not party to the NPT, and no prior consent or other conditions on spent fuel management. Competition for export of enrichment technology is also taking place between Eurodif and Urenco in negotiations for establishing an enrichment facility in Australia.

Possible future developments

This analysis has made clear that from a non-proliferation point of view existing multinational arrangements are far from ideal. To improve the situation it is necessary either to create different, more proliferation-resistant multinational arrangements, or to start from the present situation and improve existing arrangements. There is also the problem of integrating national facilities into the multinational enterprises.

An attempt to create entirely new arrangements seems unattractive for two reasons. First, it would enlarge the already existing overcapacity in enrichment services, and second, it would raise a large number of difficult questions, including membership of various countries, location of the plants, and so on [51i]. It would seem far more reasonable to adjust the present arrangements to give them more desirable non-proliferation features. However, the success of such an attempt is heavily dependent on the co-operation of the owners of existing enrichment plants and their governments, who must be convinced to sacrifice certain prerogatives in the interest of a more secure non-proliferation regime.

Table 3.1. Status of implementation of non-proliferation conditions in existing multinational arrangements

Non-proliferation question[a]	Multinational arrangements		
	Urenco	Eurodif	Euratom
Participation only of NPT parties?	Yes; not explicitly required	No	No
Governmental participation, intergovernmental decision structure for non-proliferation matters? (2)	Yes; Joint Committee (according to Almelo Treaty) responsible for non-proliferation questions	No; but 1980 Eurodif Treaty between governments contains some general non-proliferation obligations	Yes; Euratom Treaty provides decision structure (in practice high degree of independent national behaviour)
Safeguards? (1)	Yes; Euratom safeguards, verified by IAEA	No; but negotiations for Euratom safeguards under way	Yes; Euratom safeguards, in non-nuclear weapon states verified by IAEA
Withdrawal of participants excluded? (2)	No; permitted since 1980, one year after notification	No; withdrawal by negotiation and selling of share	Yes; Euratom Treaty is of unlimited duration (Article 208) and contains no withdrawal provisions[b]
Enrichment under national control prohibited? (3)	No; except for obligation to offer co-operation on centrifuge technology	No	No
Enrichment R&D only under multinational auspices? (3)	No; except for obligation to offer co-operation on development in centrifuge technology	No	No
Only one enrichment facility? (4)	No; facilities in UK and the Netherlands, one under construction in FR Germany	Yes; in nuclear weapon state	No; facilities under national or multinational control in several countries
Prohibition of transfer of sensitive technology to non-participant countries? (3)	No; negotiations on centrifuge technology transfer held and under way	No; negotiations on transfer of diffusion technology under way	No

Sharing of sensitive technology between partners excluded?	No; in principle obligation to share	Yes; in practice, but not in principle	No
Production restricted to LEU excluding production of HEU? (3)	No; but prohibition of use of HEU for nuclear explosive devices; no production of HEU as yet	No; but prohibition of use of HEU for nuclear explosive devices; no production of HEU as yet	No; production of HEU in UK and France, also for export

[a] The numbers (1), (2), (3), and (4) refer to the conditions for non-proliferation effectiveness mentioned on pp. 62–63.

[b] According to the Vienna Convention on the Law of Treaties (Article 56), this in general means that a treaty is not subject to withdrawal. However, opinions may differ about this question. It is clear that a country could make the *political* decision to withdraw even if this would violate the treaty.

Internationalization

Basic idea

The most radical institutional approach to preventing the proliferation of nuclear weapons is the creation of a single international authority, which owns and manages all or part of the nuclear fuel cycle. All countries engaged in nuclear activities would adhere to this authority.

The first internationalization proposals were made in 1945, at a time when no substantial activities for the civilian use of nuclear energy were yet under way. These proposals considered internationalization of the whole nuclear fuel cycle; that is, from uranium mining to spent fuel storage and waste management, with special attention paid to the dangerous parts. More recent proposals have been aimed at internationalization of only the sensitive parts of the nuclear fuel cycle: enrichment, reprocessing, fuel fabrication, spent fuel management, plutonium storage, mixed-oxide fuel fabrication plants and waste disposal [71]. Other parts, such as uranium mining, milling, and the operation of power reactors, are supposed to be nationally owned and operated, although it is assumed that they would be under effective international safeguards. Such a limited internationalization scheme seems easier to achieve at present, but even this proposal faces many obstacles.

Attention naturally focuses on sensitive facilities, both because these are more difficult to safeguard effectively [51j] and because they allow direct access to material which could be used to produce weapons in a relatively short time. Internationalization of the sensitive components of the nuclear fuel cycle would be a means of containing the technology and assuring that the most proliferation-prone activities of peaceful nuclear energy production were operated in a way that made access to such material much more difficult for individual states. It would also promote mutual confidence among nations by allaying fears of misuse of peaceful nuclear technology [51k].

To make an international regime successful the following features seem necessary or desirable:

1. No country should be allowed to undertake sensitive nuclear activities on a national scale [51m].

2. Research and development in sensitive fields should be performed only under the control of the authority.

3. The treaty establishing the international authority should contain no withdrawal provisions.

4. The authority should not provide any nuclear services to nations not party to the treaty. The intention of this restriction would be to provide an incentive for all nations to join the treaty. However, as has been previously noted, such restrictions can have the opposite effect of stimulating independent development of nuclear technology unless adequate pro-

visions are included in the treaty to ensure that all legitimate needs of participating states for facilities and materials will be met.

5. The authority should have the power to enact sanctions against violators [51n]. Some examples of violations which could provoke sanctions would include manufacturing and detonating nuclear explosive devices, cheating safeguard controls, withdrawing from the treaty, seizing a facility or other property belonging to the authority, and various clandestine national activities related to sensitive nuclear technology and materials. Sanctions could take many forms, but for various legal, practical and political reasons the sanctions most likely to be agreed upon would be those directed against the nuclear activities of the offending state [51o]. Actually, the existence of full internationalization of the sensitive parts of the fuel cycle would very likely make the imposition of sanctions both easier to arrange and more effective.

This requirement for sanctions is somewhat more controversial than the others. It can be criticized on at least two bases. First, it can be seen as politically unrealistic in a world in which the prerogatives of national sovereignty are jealously guarded by states, and second, it can be argued that even without explicit sanctions an international treaty can strongly inhibit violations simply by force of world opinion. There is some merit to these criticisms, and they imply that if sanctions are to be included in the treaty they will have to be designed with great care to remove any suspicion of discrimination or excessive interference in the internal affairs of states. Obviously, it is also necessary that the authority should guarantee adequate supplies of nuclear fuel to all member countries who fulfil their non-proliferation obligations.

Evaluation of past efforts

In the original 1946 Lilienthal proposal for internationalization of atomic energy, it was emphasized that "a system of inspection superimposed on *an otherwise uncontrolled exploitation of* atomic energy by national governments will not be an adequate safeguard" [31b]. The Baruch proposal, which was the version of the Lilienthal plan presented to the UN Atomic Energy Commission, proposed the creation of an international atomic development authority to which should be entrusted "all phases of the development and use of atomic energy". This authority was to set up a thorough plan for control of the field of atomic energy "through various forms of ownership, dominion, licenses, operation, inspection, research and management by competent personnel" [72a].

An important distinction was made between control and inspection [73]. The authority was supposed to have "managerial control or ownership" of all atomic activities "potentially dangerous to world security", while it should merely have "power to control, inspect and license all other atomic activities" [72b]. Thus the authority was not itself to exploit uranium resources, but it should have "complete and accurate information on world supplies of Uranium and Thorium and bring them under its dominion" [72c]. On the other hand, in order to "exercise complete managerial control of the production of fissionable materials" the authority was supposed to "control and operate all plants producing fissionable materials in dangerous quantities and to control the product of these plants" [72d] (which include enrichment plants). These intrinsically dangerous activities, as well as stockpiles of raw and fissionable materials, were to be distributed throughout the world in order to avoid appropriation by one country. Also, the actual operation of any plant containing fissionable material with the "potential of dangerous use" was to be "under the management, supervision and control of the Authority" [72e]. The Baruch proposal did not deny to countries the right of conducting research (except in explosives) under national jurisdiction. However, it intended to charge the authority with considerable research and development responsibilities in order to be "the world's leader in the field of atomic knowledge and development" [72b] .

A comparison of this proposal with the previously listed characteristics of an effective international treaty shows that the Baruch–Lilienthal Plan incorporated most of them; however, a sensitivity to the political context in which the Plan was being offered was missing. This lack of political understanding has been interpreted in many ways, ranging all the way from naïvety to cynicism, but whatever the reasons, it was clear to most people, even at the time, that the Plan would not be acceptable to those states whose endorsement was essential for its success.

Since the failure of these early attempts no serious efforts have been made to bring about full internationalization of the nuclear fuel cycle. At the present time no major internationally owned or managed facilities exist. However, in the field of inspection some internationalization has been established, in particular the safeguards authority of the IAEA which was evaluated in section II and which has been given the task of implementing the safeguards provisions as required by the NPT. Besides its provisions for safeguards, the NPT has a number of other aspects relevant to a consideration of international control of nuclear activities which should be evaluated.

Towards a universal non-proliferation regime: the NPT

A number of factors, including the undiminished nuclear arms race between the USA and the USSR, the increasing radioactive contamination caused by atmospheric nuclear test explosions, and the increasing danger of nuclear proliferation, led in 1960 to initiatives in the United Nations by a number of smaller and non-aligned countries for creating a more universal non-proliferation regime. These initiatives culminated in 1968 in the Non-Proliferation Treaty, which took effect in 1970. This Treaty was the result of many years of negotiations in which the USA and the USSR at last recognized their common interest in non-proliferation matters [74a]. By that time the number of countries possessing nuclear weapons had already grown to five.

The NPT makes a fundamental distinction between 'nuclear weapon states', defined as states which have "manufactured and exploded a nuclear weapon or other nuclear explosive device prior to January 1, 1967", and 'non-nuclear weapon states'. The Treaty includes the following three major commitments by its parties.

1. Non-nuclear weapon states party to the Treaty undertake not to acquire in any way the possession of nuclear weapons or other nuclear explosive devices, not to acquire control over such weapons or explosive devices directly, or indirectly. For their part the nuclear weapon states party to the Treaty will in no way give any assistance to such attempts, (Articles I and II).

2. "Each of the Parties to the Treaty undertakes to pursue negotiations in good faith on effective measures relating to cessation of the nuclear arms race at an early date and to nuclear disarmament . . . " (Article VI).

3. The peaceful use of nuclear energy shall be stimulated, "especially in the territories of non-nuclear-weapon States Party to the Treaty, with due consideration for the needs of the developing areas of the world". All parties to the Treaty have the right to participate in "the fullest possible exchange of equipment, materials and scientific and technological information for the peaceful uses of nuclear energy" (Article IV). At the same time an international safeguarding system has to verify that in the non-nuclear weapon states no special material will be diverted to manufacturing nuclear explosive devices (Article III).

The basic hope of the NPT was that it would be possible to create an adequate barrier between peaceful and military use of nuclear energy, and that this separation could be effectively realized by means of an international safeguarding system. The NPT assigned the application of these safeguards to the International Atomic Energy Agency, which had previously been proposed by President Eisenhower in the 1953 'Atoms for Peace' speech and was established in 1957. It must be emphasized that the NPT does not require any international ownership of nuclear facilities or material. All nuclear activities are in principle under national control. Only the safeguarding system has been internationalized.

The NPT was aimed at establishing an effective universal non-proliferation regime, one which would encompass all peaceful nuclear activities and to which all countries would submit. To be successful it should have been the result of a universal international consensus on the world's nuclear weapon and proliferation problem; but consensus was not reached, and several major countries engaged in nuclear activities refused to become parties to the Treaty.

The objection most often raised against the NPT is that the Treaty is fundamentally discriminatory. It discriminates in its privileges and obligations between nuclear weapon states and non-nuclear weapon states. Certainly the feature that is most obviously discriminatory is its failure to prohibit the possession of nuclear weapons by the nuclear weapon states. The call of Article VI for cessation of the nuclear arms race is formulated in very general terms, without commitments to any specific measures. In addition the NPT does not contain any provisions of security guarantees to the non-nuclear weapon states against nuclear aggression, nor sanctions for violators of the Treaty.

The entry into force of the NPT in 1970 brought about no diminution of the nuclear arms race, the so-called 'vertical' proliferation of nuclear weapons. At the First Review Conference of the NPT in 1975 it was evident to many non-nuclear weapon states that the nuclear weapon states party to the Treaty (the USA, the USSR and the UK) had not taken sufficient measures to halt vertical proliferation. Nor were they prepared to make any commitment to specific measures to that effect, such as a Comprehensive Test Ban, as proposed by the non-aligned states. It was clear that the NPT's discriminatory character remained the main source of contention between the non-aligned states and the nuclear weapon states.

The NPT is also discriminatory with regard to the application of safeguards. These are not required for the peaceful nuclear activities of the nuclear weapon states party to the Treaty. Non-nuclear weapon states consequently saw themselves in a disadvantageous position because their nuclear industries were more susceptible to possible industrial espionage by IAEA inspectors. France, the UK and the USA did announce in 1967 that they were prepared to bring their civilian nuclear facilities under IAEA safeguards. However, much remains to be done to implement these assurances. For example, an agreement between the USA and the IAEA on safeguarding US facilities was not ratified until autumn 1980 and specific arrangements for inspection are still far from determined.

Another discriminatory element in the NPT's application of safeguards is the implied distinction between two types of non-nuclear weapon states: those who are parties to the Treaty and those who are not. Whereas the former are obliged to have *all* their peaceful nuclear installations and materials under IAEA safeguards (a full-scope safeguards requirement), the latter are required to submit to IAEA safeguards only the nuclear material and installation which are supplied by countries party to the Treaty. No safeguards are required on nuclear materials and installations acquired indigenously or from countries not party to the Treaty. This aspect acts as a disincentive for countries to become party to the NPT. Even though a non-party state may be in the same position as an NPT country *vis-à-vis* supplier states, at the same time it can put itself in a more advantageous position with respect to safeguards inspection by staying outside the NPT. At the 1975 Review Conference nuclear supplier countries still refused to agree to require full-scope safeguards for non-nuclear weapon states not party to the Treaty, although a few years later a number of supplier states (Australia, Canada, Sweden and the USA) unilaterally took this step.

The several discriminatory aspects of the NPT just described are certainly present in the Treaty, but they do not provide a sufficient explanation for a state to refuse to become a party to the Treaty. It must be emphasized that many non-nuclear weapon states have willingly accepted these restrictions, and it must also be assumed that this was done in the belief that joining the NPT was in the state's best interests. In other words, many states have come to the conclusion that it is worthwhile to sacrifice certain aspects of national sovereignty in the interests of long-term security. Given this choice by the great majority of nations in the world, it becomes more difficult to accept at face value the claims of discrimination made by a relatively small minority.

Limitations of the NPT

Because no universal international consensus was reached on the NPT, the scope of the NPT non-proliferation regime is seriously limited. For example, two nuclear weapon states, France and the People's Republic of China, did not join the Treaty, although France has declared that it would "behave in the future in this field exactly as the States adhering to the Treaty" [74b]. China in practice has acted likewise. As far as can be judged, none of the parties to the NPT, 116 in number by 1982, have violated the Treaty, although Iraq's temporary blocking of the inspection of its nuclear installations in 1980, in connection with the Iraqi–Iranian war, came close to it. Many also consider the nuclear arms race as a lack of good-faith efforts by the nuclear weapon states towards halting vertical proliferation as required by Article VI. Moreover, it is important to note that several non-nuclear weapon states, with both substantial nuclear programmes and chronic national security problems, have not joined the NPT. Among these, India exploded a nuclear device in 1974 based on plutonium produced in a Canadian-supplied research reactor using heavy water supplied by the USA. India thus became the first manifest proliferation case after the NPT had entered into force in 1970. Israel has been reported to possess several nuclear bombs since the Yom Kippur War of 1973 (or to be able to assemble these bombs in a very short time) [74c, 75]. South Africa has been reported to have prepared a nuclear test in the Kalahari desert in 1977, allegedly cancelled under US pressure [76b]. The South Africans have also been suspected of exploding nuclear devices somewhere in the Southern Atlantic or Pacific area in 1979 and 1980, but there is still some dispute over the precise nature of and responsibility for these events [77, 78, 79].

The difficulties faced by the NPT are best summarized by the results of the Second Review Conference in September of 1980 (see chapter 7).

Just as at the First Review Conference one of the central issues was the failure of the nuclear weapon states to fulfil their obligations to good-faith negotiations towards halting the nuclear arms race at an early date. Very little progress in this area had been made in the period between the First and Second Review Conferences. In particular, no Comprehensive Test Ban had yet been concluded, and the Strategic Arms Limitation Talks (SALT) II agreement had not been ratified. On the contrary, vertical proliferation was accelerating, as illustrated by the development and scheduled introduction of several new types of nuclear weapon systems.

Because of these disagreements the Second NPT Review Conference was not even able to produce a Final Declaration, something which had at least been achieved at the previous review. The most alarming development seems to be the increasing interest of the nuclear weapon states in the idea of fighting a limited war using nuclear weapons. If such developments continue, they can only increase the incentives for non-nuclear weapon states to re-evaluate their decision not to acquire nuclear weapons.

In conclusion it can be said that although the NPT represents a genuinely positive development in the effort to establish a peaceful international nuclear energy regime, it still falls far short of what is necessary to ensure that nuclear energy will be employed only for peaceful purposes. Only in the area of safeguards has the Treaty established a functioning international body, and even this safeguards system does not cover many of the nuclear facilities in the world. However, even if safeguards were extended to all nuclear facilities, this would represent only a partial step towards preventing nuclear weapon proliferation. This was understood from the earliest days of the original Lilienthal proposal [31b].

International Nuclear Fuel Cycle Evaluation (INFCE)

One other significant attempt at collaboration was the two-year study carried out by INFCE. The scope and purposes of this study are described in detail in chapter 7, so they will be stated only very briefly here. The major purpose of the study was to establish the technical basis for possible future non-proliferation measures. The study was divided into a number of Working Groups, one of which concerned itself with the issues of uranium enrichment and fuel supply.

In contrast to the NPT, INFCE dealt only with the problem of horizontal proliferation resulting from an increasing use of nuclear energy. Moreover, it considered only the technical aspects of this problem. INFCE did not consider how attempts to prevent horizontal proliferation might be negatively influenced by the continuing vertical proliferation, although it was recognized at the Organizing Conference of INFCE in October 1977 that "a decision by a government to acquire nuclear weapons is essentially a political decision motivated by political and national security considerations, among which is the relationship between vertical and horizontal proliferation and the existing and undiminished arms race" [13i].

Thus in its desire to stimulate the use of nuclear energy on the one hand and to prevent the spread of nuclear weapons on the other, INFCE faced essentially the same dilemma as the NPT. However, by defining its objectives in purely technical terms, INFCE managed largely to avoid the more difficult political issues. But, even technical analyses and discussions may influence politics, and it is possible that some useful political results may yet follow from these technical deliberations. One positive result has been that non-NPT countries, such as Argentina, Brazil, France and India, have participated in the discussions.

The political negotiations which follow INFCE will undoubtedly focus on the issue of restriction of access to weapon-grade material by limiting the spread of sensitive technology and facilities. The negotiations could result in an agreement among technologically advanced countries not to transfer sensitive technology to less advanced countries, but to retain the right to exploit reprocessing and enrichment facilities under national control. Such an agreement would alter the present distinction between nuclear weapon and non-nuclear weapon states to one based more on the state's degree of technical advancement. Technically less advanced states would then be dependent for their nuclear fuel supplies either on other countries or on some institution such as an international fuel bank, which would necessarily be provisioned by the advanced countries.

For the advanced countries, which would possess enrichment or other sensitive facilities, the technological threshold for obtaining weapon-grade material would be considerably lowered. The dividing line between countries with potential physical access to weapon-grade material and countries without such access would consequently be shifted roughly from a position between the advanced and middle developed countries to a position between the middle and less developed countries. Such a policy would enhance the possibilities of middle developed countries manufacturing a nuclear bomb in a relatively short time. Such an arrangement would at best be capable only of delaying the proliferation of nuclear weapons to less developed countries.

Prospects for further internationalization

It has been argued that the time may be ripe for a more determined approach to internationalization [51p]. Along these lines the INFCE conference recognized in its report that there are no fool-proof technical solutions to the proliferation problem and stressed the need for further institutional measures [13j]. More specifically, the USA in its Nuclear Non-Proliferation Act of 1978 called for the establishment of an International Nuclear Fuel Authority (INFA). Such an authority would assure fuel supplies to countries under non-proliferation conditions consistent with the following provisions: that the recipient country (*a*) accepts full-scope IAEA safeguards, (*b*) does not manufacture or otherwise acquire any nuclear explosive device, (*c*) does not establish any new enrichment or reprocessing facility under its national control, and (*d*) places any such existing facilities under effective international auspices and inspection [80]. However, this proposal has generated little support from other states. There has also been little progress reported from the three IAEA Working Groups studying international measures for plutonium storage, spent fuel management, and fuel supply.

Any attempt to establish internationally owned and managed enrichment facilities poses the same dilemma as in the multinational question; that is, should the system be formed from existing facilities or should it establish a new set? Here the case for internationalizing existing facilities seems even stronger than in the multinational situation. Only in this way could the result be a single international enrichment service authority, thereby diminishing the danger that its enrichment supply policies could be undermined by independent national facilities.

However, many complex issues remain to be resolved before such a fully internationalized system could be achieved. These include economic questions concerning the pricing of fuel and the distribution of costs among the member states of the international system, and political questions regarding conditions for the supply of enrichment services. Obviously, the greatest political question of all is whether there is even sufficient interest in the international community in such an arrangement to give it some hope of success.

Chapter 4. Conclusions and recommendations

I. Introduction

Following this introductory section, this chapter is divided into two parts. First, in section II a number of conclusions are drawn from the analysis of chapters 1–3. These conclusions summarize our assessment of the degree to which developments in the uranium enrichment industry are stimulating or facilitating the horizontal proliferation of nuclear weapons, as well as our judgement on how effective existing control mechanisms are likely to be in limiting this process. Then in section III we recommend a number of measures that could improve the situation.

Before listing our conclusions and recommendations it would be helpful to make clear two important assumptions which underlie everything that follows. The first is that the horizontal proliferation of nuclear weapons is undesirable and worth considerable effort to prevent; and the second is that the use of nuclear energy for non-military purposes will continue in the foreseeable future.

The first assumption, although shared by many individuals and states, is neither provable nor universally accepted. In the late 1950s and early 1960s French nuclear strategists argued that the proliferation of nuclear weapons and their attendant 'deterrent' capabilities would actually contribute to world stability [81]. This argument has recently been revived by a respected US analyst in a thoughtful and carefully reasoned essay [82]. The basic point of this argument is that the slow spread of nuclear weapon capabilities to more nations can have positive effects and is in any event not worth the heroic and politically difficult efforts which will be necessary to prevent it.

An attempt to refute these arguments is unfortunately beyond the scope of this book. It must suffice to say that we do not find the arguments persuasive and regard the further horizontal proliferation of nuclear weapons as one of the major dangers to world peace and security. At the same time we recognize that the problem must be kept in perspective and not allowed to be used either to justify unacceptable actions by one state against another or to obscure the far more dangerous implications of the vertical proliferation currently being accelerated by the USA and the USSR.

We have made clear in chapter 3 (see p. 55) our belief that the dangers of nuclear proliferation do not justify the unilateral decision by one state to attack the nuclear facilities of another, and it is important to emphasize here our conviction that the nuclear arms race between the USA and the USSR is not only a greater threat to world security than horizontal proliferation, but must be seen as one of the primary factors stimulating the further spread of nuclear weapons. The minimal attention paid to vertical proliferation in this book should be regarded as the result of a lack of space rather than a lack of concern.

Our second major assumption is that the use of nuclear energy for peaceful purposes will continue at some level of activity for the foreseeable future, and that any efforts to prevent the spread of nuclear weapons must be carried out within this context. There are some who would argue that this assumption is tantamount to accepting the inevitable further spread of nuclear weapons to more countries. From this point of view the commercial and military applications of nuclear energy are so closely linked that any attempt to separate capabilities from intentions is doomed to failure [83a]. With this assumption it is possible to argue logically that a necessary condition for the prevention of nuclear proliferation is the elimination of nuclear power. In this view all nuclear activities are inherently dangerous (there is no such thing as 'peaceful' nuclear power), and when this is coupled with the belief that nuclear energy is both politically and economically disastrous anyway [83b], a powerful argument emerges for the abandonment of all nuclear technology.

However, this argument rests heavily on the assumption that capabilities and intentions cannot be separated, an essentially unprovable hypothesis. Some countries which have developed the full range of nuclear capabilities have so far chosen to renounce nuclear weapons. These nations, for a variety of reasons, have concluded that it is not in their interests to produce such weapons even though they could do so if they wished. This historical evidence undermines the assumption that the spread of nuclear energy technology makes inevitable the spread of nuclear weapons. This connection seems no more inevitable than, for example, the expectation that the spread of industrial chemical technology will lead inevitably to the acquisition of chemical warfare arsenals by all nations. Such assumptions represent a kind of technological determinism, and a very pessimistic kind at that, which ignores the influence of political and social factors which might remove the motivations to apply technology to military uses. The almost universal rejection of the use of chemical and biological warfare stands as a counter-example to this fatalistic assumption.

Therefore, a recommendation to unconditionally abandon nuclear power, without paying attention to the role of nuclear weapons in national defence and international policies, will not provide a solution to the problem of nuclear weapon proliferation. However, we note with approval the clear evidence that more nations, both developed and developing, are taking a much more sceptical and realistic look at nuclear energy than was done in the past. The present malaise in the nuclear industry shows that the unallayed optimism and boosterism of the 1960s are over, and that any future use of nuclear technology will be based on a far more conservative assessment of its costs and benefits. This particularly applies to developing countries, many of whom are realizing that the enormous capital demands, administrative centralization and technological sophistication of nuclear processes are especially difficult to reconcile with desires for broad-based economic growth and lowered dependence on foreign economic and technical assistance. Since it is largely the developing countries which are most often identified as candidates for proliferation, this reassessment of civilian nuclear energy seems to make less likely the possibility that widespread, intensive use of nuclear energy in these countries will make weapon proliferation more difficult to control.

The problem of nuclear weapon proliferation will not be solved until the root cause is attacked. This is the belief that nuclear weapons are desirable things to have, whether for use in fighting wars, in deterring and/or intimidating adversaries, or only as symbols of technological power or political equality with the mighty. Until all these motivations are shown to be empty, and until nuclear weapons are outlawed by a world consensus that they are too dangerous, politically and militarily counter-productive and morally unacceptable, the threat of nuclear proliferation will remain.

It is not necessary to point out how far the world is from such a consensus. Indeed, these principles seem to be less widely believed today than they were in the aftermath of the destruction of Hiroshima and Nagasaki. Despite the fact that nuclear weapons "have indiscriminate effects *per se*, and are inhumane, cruel and repulsive" [84], they have not been outlawed, only forbidden in a blatantly discriminatory manner to a limited number of nations in return for what remain empty promises of general nuclear disarmament. Meanwhile, the two great powers continue to escalate the arms race and to devise new weapons and the strategies which rationalize them.

The conclusions and recommendations which follow must be read in this context. In the present environment these measures can at best provide partial and temporary results in slowing further proliferation. They affect only capabilities, not motivations. Given a genuine commitment to nuclear disarmament these recommendations could provide time to work on reducing these motivations before nuclear weapons spread to many more countries. Without such a commitment, the extra breathing time will be of no ultimate value.

II. Conclusions

1. *Technological thresholds to proliferation via the enrichment route have decreased significantly in recent years and are in danger of decreasing even further.*

The variety, efficiency and accessibility of uranium enrichment technology have improved substantially since the early 1960s. The strong commercial motivations which characterized the period of growth and optimism in nuclear energy led to a heavy investment in research and development in enrichment processes which would be more compact, flexible and energy-efficient than gaseous diffusion. Unfortunately, these same characteristics would be advantageous to any country that wishes either to adapt an existing facility or construct a dedicated facility for the production of highly enriched, weapon-grade uranium.

This problem is best exemplified by the gas centrifuge, a technique which represents a qualitative advance over gaseous diffusion in the above respects. This has resulted in its commercial adoption by the three Urenco countries, in a scheduled commercial plant in the USA, in an intensive R&D programme in Japan, and in its clandestine appropriation by Pakistan for a facility which seems to have unmistakable military implications. As centrifuges continue to be improved the already alarming potential for small, clandestine facilities will increase, especially if competitive pressures for exports of technology also increase (see conclusion 3 below).

Meanwhile, the next generation of enrichment processes, in particular the molecular laser and plasma techniques, are undergoing development. If these are successful, they promise to accentuate even further the dangerous trend towards proliferation-prone methods. Most worrisome is the molecular laser process, which promises exceptionally low specific energy requirements, high single-stage enrichment, rapid product collection and reflux and very short production times. If it can be shown that such performance characteristics are achievable at reasonable cost and with accessible technology, and the effort to demonstrate this is going on in many countries, then another major step will have been taken towards making national enrichment capabilities accessible to a much wider range of users and potential abusers. One cannot even exclude the possibility of subnational groups acquiring such capabilities, although at the present time this seems unlikely even under pessimistic assumptions.

2. *Incentives for acquisition of national enrichment capabilities remain high, reflecting a continuing interest in energy independence, resource development and/or a nuclear weapon option.*

The analysis of chapter 2 shows a significant lowering of both technological and situational thresholds to proliferation via the enrichment route. There is still a high premium on national energy independence, and

84

when this is added to the persistent belief in many countries that, besides coal, nuclear energy is the only viable alternative to oil, the arguments for a national enrichment capability seem strong. Such arguments can only be valid for countries with guaranteed access to uranium resources, and even for these countries nuclear energy still presents economic, political and environmental problems which make its widespread application much less attractive than it once seemed. However, national desires for energy independence remain strong, and nuclear energy has substantial technological and bureaucratic momentum as well as an international agency firmly committed to its promotion. This ensures that, whatever its problems, it will continue to be a part of the world energy picture for some time to come.

Coupled with this push for energy independence is a continuing high level of international tension in many areas of the world, for example, in southern Asia, the Middle East and southern Africa. In such areas war is an ever-present danger, and it is not surprising that nuclear weapons can appear attractive to national leaders as deterrents against attacks or threats by rival states; but the experience of the past 35 years has shown that such deterrent postures are most likely to be interpreted by rivals as threatening and are therefore inherently destabilizing.

Unfortunately this lesson has still not been learned. The nuclear weapon states have expressed their own confidence in the value of nuclear weapons in the most unambiguous way, by investing heavily in new generations of usable nuclear weapons. In the process they have made a mockery of their NPT commitment to good-faith efforts at nuclear disarmament and continue to set an example which can only make nuclear weapons more attractive to countries who face potential regional or even domestic enemies.

3. *A buyer's market exists for enrichment services making it difficult to maintain a consensus on safeguards requirements among suppliers.*

The many years of optimism over the future growth of nuclear energy led to large investments in research, development and operating facilities in national enrichment industries. The recent stagnation in nuclear energy growth has resulted in an over-supply of enrichment capacity and an intense competition for markets among the supplier countries. This competition has in some cases taken the form of showing a willingness to relax safeguards requirements on transferred materials and enrichment services in an attempt to gain a competitive edge.

With respect to the supply of enrichment services to Brazil, for example, FR Germany and the UK were prepared to demand less strict safeguards than their Dutch Urenco partner, which led to strained relationships among these countries. In addition to supplying nuclear material and services on the international market, some countries have also tried to become major exporters of nuclear technology and installations. This desire often had strong domestic motivations, such as the possible exchange of technology for raw materials, an attempt to maintain a viable nuclear industry, or the desire to create or sustain a profitable trade balance. The combination of strong national desires for sales and the lack of a solid international consensus on safeguards has led to some questionable deals, notably the West German sale of an entire fuel cycle, including an enrichment plant, to Brazil. In the London Club the major nuclear exporters have reached some agreement on restraints and the demand for safeguards on nuclear exports, although competitive pressures for sales of components and equipment have also led to disagreements among members.

A possible future development, which may affect the enrichment market, is that countries with large uranium resources will seek to export this uranium in enriched form. Some of these — either independently (South Africa), or in co-operation with others (Brazil) — are developing their own technology. In addition, Australia is seriously seeking co-operation in the field of enrichment or trying to buy the technology for it, and there have been contacts with Eurodif, Japan and Urenco. Australia could possibly either independently build a complete enrichment facility or acquire one by joining one of the existing multinational enrichment companies.[1] The latter could also be profitable for one of the existing companies as Australian membership would ease their linking together the sale of uranium ore with enrichment services.

Regardless of what form such an arrangement takes, any new enrichment capability would exacerbate the current over-supply of enrichment services and increase the need to find customers for them. All the evidence points to a continuation of the buyer's market for enrichment services at least into the 1990s, a situation that has serious implications for non-proliferation efforts.

4. *The Non-Proliferation Treaty and IAEA safeguards systems, while certainly important, suffer from serious weaknesses and contradictions which limit their ability to prevent proliferation.*

The necessity for an international treaty, such as the NPT, and a safeguards system, such as the one administered by the IAEA, cannot be denied. If the world ever does succeed in eliminating nuclear weapons, such measures will be an integral part of preventing their re-emergence. However, in the present world these measures suffer from serious deficiences, and these deficiencies seem to be growing more important with the evolution of enrichment technology.

Probably the most serious and fundamental flaw in the NPT regime is its explicit discrimination between nuclear and non-nuclear weapon states and the highly unequal distribution of obligations and privileges between these two groups. Such discrimination may be tolerable in the short run if

---

[1]   As this book was going to print the Australian government announced that it had selected Urenco for a joint venture to help Australia build a centrifuge enrichment plant and develop a uranium enrichment industry [85].

higher objectives are at stake and progress towards their attainment is visible, but in the long run it only reinforces the idea that possession of nuclear weapons carries with it a special kind of power, prestige and freedom of action. So the longer the NPT remains in effect without nuclear disarmament by the nuclear weapon states, the greater will be the forces tending to destroy the Treaty. Such forces have been increasingly evident at the two Review Conferences which have been held since the original signing of the Treaty.

This asymmetry of the NPT has been cited as the major reason for refusing to sign it by most of the countries who remain outside the Treaty. These countries generally have not accepted full-scope IAEA safeguards for all of their nuclear facilities, although many have accepted limited safeguards in order to obtain technology and fuel supplies from abroad. The result has been an uneven and inadequate application of safeguards to nuclear facilities in general and almost none to enrichment plants in particular.

The technology of enrichment plant safeguards is in a relatively primitive state, and IAEA standards for timely detection of significant diversions are demonstrably inadequate in today's technological and political environment. Even at safeguarded enrichment facilities the operators are allowed to declare the cascade area off-limits to inspectors and to restrict containment and surveillance measures to only the most non-intrusive and therefore ineffectual level. Although there are some ideas on how these conflicting demands can be reconciled, none of these so far offer much hope of providing genuine reassurance that diversions can be detected.

The situation is complicated further by suggestions that enrichment processes such as chemical exchange should be encouraged as an anti-proliferation measure. These processes, by virtue of their long equilibrium times and large uranium inventories, promise to be far less susceptible to modification or reprogramming for producing weapon-grade uranium. But these same qualities compound the difficulties of applying materials accountancy methods to detect diversions of low-enriched product, which could then be further enriched in a much smaller, possibly clandestine, facility. This problem underlines the conclusion that no enrichment process is proliferation-proof; each one seems to have its own particular weaknesses and dangers.

5. *Multinational institutional measures and arrangements show some promise of retarding proliferation, but the existing enrichment consortia, Urenco and Eurodif, and Euratom all have serious deficiencies which make them less effective than they might be.*

The formation of Urenco and Eurodif was not motivated primarily, or even significantly, by non-proliferation objectives. Consequently they have serious weaknesses from the perspective of controlling proliferation. Both consortia are interested in selling enrichment services on the world market in addition to supplying their own enrichment needs. This leads to the same kind of dangerous competition as would be expected from purely national facilities (see conclusion 3 above).

In addition to this common weakness, each arrangement has its own particular flaws. Urenco's policy of placing enrichment facilities on the territories of each of its members, its withdrawal clause and its difficulties in agreeing on the scope of safeguards requirements all weaken its ability to enforce non-proliferation objectives. On the other hand, the exclusive control of Eurodif policy-making by the French CEA obviates many of the advantages of such a multinational arrangement.

Moreover, the Euratom Treaty is basically only a paper arrangement. Except for safeguarding, it has in practice no real control over the enrichment activities of its member states.

III. Recommendations

1. *The enrichment industry should be internationalized, possibly along the lines of an international nuclear fuel agency (INFA)*[44b].

All national enrichment facilities should be brought under the authority of this agency, which would own and operate them in response to national demands for enrichment services. Such an agency would be responsible for the production, distribution and safeguarding of enriched uranium, but would have no mandate to promote the use of nuclear energy (see recommendation 2 below). Supply of adequate amounts of enriched uranium would be guaranteed to all members of the agency in return for acceptance of full-scope safeguards (see recommendation 4 below).

Since the total capacity of the world's enrichment plants is considerably in excess of current and projected demand, a number of planned facilities, such as the US plant in Portsmouth, Ohio and the planned Urenco additions at Gronau, FR Germany, Almelo, the Netherlands and Capenhurst, UK, should be deferred.

2. *All research and development on uranium enrichment should be conducted by INFA, and efforts to develop the molecular laser enrichment method should be stopped.*

The potential savings in energy and capital offered by this process are not worth the substantial extra proliferation threat it would create. It should be emphasized that such a decision would not interfere with basic and applied research in the scientifically interesting and potentially useful fields of laser photochemistry and isotope separation. Only the development of laser enrichment of uranium would be discouraged. Such a distinction is possible in principle because of the very specialized properties of the lasers and materials handling systems required for uranium enrichment.

These same comments apply to the various plasma separation processes currently under study. These development programmes should also be stopped unless it can be shown that some plasma processes possess features which make them unusable for the separation of uranium and plutonium isotopes.

It is easy to make such a recommendation, but much harder to visualize its implementation. If independent research and development of certain processes is to be prevented, then the question naturally arises as to how this is to be carried out. In particular, the recommendation can be interpreted as suggesting an elaborate and intrusive inspection mechanism to detect violations.

We recognize the severe political and administrative problems raised by such a mechanism and would prefer to rely on other means to discourage such activities. In particular, the elimination of international competition in the supply of enrichment services should remove much of the incentive for developing the new processes. When this is coupled with the very high expense and commitment of scientific resources needed to develop the laser or plasma methods, it seems unlikely that very many states will choose to make such an effort. And there remains the substantial risk that the effort will be found out even without a formal inspection mechanism, exposing the state to criticism or even sanctions.

3. *The processes employed by the INFA should be either gaseous diffusion or the newer chemical-exchange methods.*

Since INFA enrichment plants would have international technical and management staffs, it would be impractical to attempt to keep most technical details of the processes secret. Therefore, it would be necessary to employ technology which presents serious obstacles to the construction of dedicated facilities above and beyond the problem of acquiring classified data. The trade-off may be some loss of efficiency and somewhat higher SWU costs in exchange for a considerable raising of the technological threshold to proliferation.

Unfortunately, the centrifuge cat is already partially out of the bag, and a number of operating facilities already exist. Preferably, these facilities should be shut down and dismantled. The US calutrons of World War II provide a historical precedent for this action, although the situations are not fully analogous. The modern centrifuge is far more efficient and reliable that the old calutrons, so it will be much more difficult to stop its use for uranium enrichment.

If it should prove impractical or impossible to shut down the centrifuge plants, then the internationalized centrifuge facilities should be managed in such a way as to prevent the further dissemination of this process. This can be done by allowing the current managers and technicians to remain at such plants, but subjecting the plants to much more stringent safeguarding than is now contemplated (see recommendation 4 below). Eventually, the objective should be to phase out the gas centrifuge technique for uranium enrichment.

The problem of effectively preventing exports of enrichment technology remains a difficult one. It is hard to imagine a politically acceptable mechanism by which an international agency could control exports of components or know-how. Probably the best that can be done is for the INFA to establish a central data bank on exports of sensitive technology and components, and other 'grey area' items, so that all data available from open sources were assembled in one place. This would facilitate the detection of patterns of exports and imports which were suggestive of attempts to develop independent enrichment facilities. Detection of such activities should trigger a warning of possible sanctions to the nation involved (analogous to the provision in the US International Security Assistance Act; see chapter 3, section III).

4. *No new national enrichment facilities should be built.*
The incentives for national enrichment capabilities can be removed by the following three strategies:

(*i*) Guarantee access to enriched fuel at reasonable prices to all countries who need it (see recommendation 1 above).

(*ii*) Encourage the satisfaction of energy needs by means other than nuclear reactors. The current situation in which nuclear energy is the only energy source promoted by an international agency should be altered to create a more balanced and comprehensive "World Energy Organization" [44c] devoted to finding appropriate solutions to the energy problems of developing and developed countries. Such an agency would certainly not exclude nuclear energy, but would place it in a much more balanced and realistic relationship with other alternatives.

(*iii*) Devalue nuclear weapons by making it clear that they are militarily useless, morally unacceptable and politically self-defeating. This can be accomplished only if the states now possessing nuclear weapons will agree to eliminate them. Then a general outlawing of such weapons can be applied equally to all states, and sanctions for violations can be administered in a just manner.

5. *Membership of an INFA, with its attendant renunciation of nuclear weapons and obligations for accepting safeguards, should be required before any enrichment services or fuel supplies are provided to a state. Membership of an INFA should not be subject to a withdrawal provision, and the authority should have the power to enact sanctions against states which either violate their agreements or withdraw from the agency.*
Such an arrangement will clearly take some time to achieve, and in the interim the existing multinational consortia, Eurodif and Urenco, should be substantially strengthened against proliferation by adoption of the non-proliferation features listed in table 3.1 (p. 70). In the highly unlikely event that more enrichment capacity is needed before an INFA can be created, such capacity should preferably be kept under multinational control in preparation for the transition to international control.

6. *Technical and administrative aspects of safeguards on enrichment facilities should be improved substantially.*

The present standard for materials accountancy accuracy of 0.2 per cent is both inadequate for the detection of significant diversions at large facilities, and unattainable for plants with large cascade inventories. This implies that even improved materials accountancy techniques will have to be supplemented by much more effective containment and surveillance measures than are now being applied. At a minimum this should include the presence of full-time international inspectors at all enrichment plants and provisions for access by inspectors to the cascade area. Furthermore, criteria such as 25 kg being defined as a significant quantity of ^{235}U in LEU and one year being the conversion time for converting nuclear materials to the metallic components of a nuclear explosive device should be re-evaluated as soon as possible. The emergence of new enrichment techniques, notably the gas centrifuge, has made these criteria seriously inadequate.

The above recommendations are derived from the conclusions of section II and are consistent with the assumptions discussed in section I. In reading these recommendations it is also necessary to keep in mind that they deal with only one aspect of the proliferation problem: the uranium-enrichment industry. Since other, equally important, paths to proliferation exist, our recommendations can make no claim to being a complete solution to the proliferation problem. At best they represent only partial and temporary measures which could begin the process of bringing the spread of nuclear weapon capabilities under control. As was pointed out in section I, a far more comprehensive and radical approach is needed to provide a stable, long-term non-proliferation regime.

Part
Two

Chapter 5. General principles of uranium enrichment

I. Introduction

Part Two is devoted to a technical description and analysis of the general features of uranium enrichment and the types of technology that have been created to implement it. The purpose of this relatively detailed treatment is to provide policy analysts and policy makers with a quantitative framework for analysing present and future developments in the enrichment industry with regard to their implications for the problem of nuclear weapon proliferation.

In Part One a number of comparisons were made and conclusions drawn concerning the relative sensitivity of various enrichment processes to a number of possible methods for producing weapon-grade uranium or diverting reactor-grade uranium for further clandestine enrichment. The descriptions in Part Two form the detailed data base for those comparisons and conclusions. It is possible that other analysts might draw different conclusions from these data, but the data themselves are intended to be as accurate and non-controversial as possible under the existing restrictions imposed by both governmental and commercial secrecy.

Although the material discussed in this Part is technical, the treatment is intended to make the material accessible to people with a relatively elementary background in the physical sciences, in particular physics and chemistry. For this reason the technical specialist in the field may find the explanations over-simplified. However, every attempt has been made to ensure the scientific accuracy of the descriptions, and to show clearly how the properties of various types of technology derive from, and are limited by, the basic physical principles which govern their operation.

The basic principles of isotope separation, the early history of the uranium enrichment industry, and short, qualitative descriptions of enrichment processes have already been given in Part One (see chapters 1 and 2). This material will not be repeated here, so the reader who would like a quick introduction to the subject is referred to the relevant sections of Part One. This Part will proceed immediately to a more quantitative approach, beginning in this chapter with a discussion of the basic concepts shared by all types of enrichment technology. These include the concept of separative work, the design of cascades, and the properties of uranium hexafluoride and metallic uranium vapour.

Chapter 6 treats each significant enrichment process separately and shows how the properties important for proliferation considerations can be either derived from basic principles, extracted from the literature, or estimated by inference from available information. No secret or proprietary information has been used in the research for this section.

The descriptions in chapter 6 emphasize those properties of the processes that are relevant to proliferation concerns. These properties are, among others, the stage separation factor, the stage hold-up time, the specific energy consumption and the status of development of the process. These properties are summarized for the 10 most prominent processes in table 6.3 (p. 188), where they can be compared conveniently. It is this table which serves as the primary source of information for the analysis of chapter 6.

One other useful result of this Part is the creation of a mathematical procedure for estimating the properties of a cascade of any desired capacity using any of the considered processes. This procedure should prove useful to analysts who wish to explore for themselves the implications of various processes or to acquire a better feeling for the sizes and amounts of equipment and material involved in uranium enrichment. Such an intuitive understanding of the enrichment technology is of great help in evaluating the significance of new developments as they occur. The mathematical procedure utilizes tables 5.1 and 6.3 and is described at the end of chapter 6 along with an example of its use.

II. Separation elements

Basic definitions

A generalized enrichment element can be treated as a 'black box' into which flows material of a certain isotopic composition and out of which flow two streams, one containing a higher percentage and the other a lower percentage of the desired isotope than was present in the feed stream. The material with the higher percentage is generally called 'product', and that with the lower percentage is called 'waste' or 'tails'. It should be kept in mind that the words 'feed', 'product' and 'tails' can be used to refer to the inputs and outputs of either a single element or a full cascade. In this section they are interpreted in the former sense.

Figure 5.1 shows a schematic enrichment element and displays the important parameters. The symbols F, P and W denote feed, product and tails flow rates and are usually expressed in units such as kilograms per second (kg/s) or tonnes per year (t/yr). Generally this flow rate refers only to the uranium content of the material, so if UF_6 is the working medium the actual mass of gas transported will be larger. Focusing on only uranium flows makes the treatment more generally applicable.

Figure 5.1. An enrichment element

Schematic of an enrichment element showing the input (feed=F) and outputs (product=P, tails=W). N_F, N_P, N_W denote the fraction of ^{235}U present in each stream.

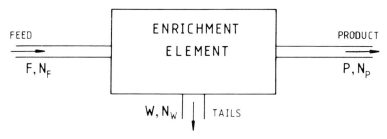

The symbols N_F, N_P and N_W refer to the percentage composition in numbers of molecules of the desired isotope (i.e., ^{235}U) in the respective flow stream. If the mixture contains only two isotopes, then the fraction of isotopes which are of the undesired type (^{238}U) will be $(1-N)$. For example, if the input to the element is one per cent ^{235}U, then $N_F = 0.01$ and $(1-N_F) = 0.99$. If the feed material is natural uranium, then $N_F = 0.0072$, which results in a percentage of 0.711 by weight [1a].

Often it is convenient to use another variable than N to describe the composition of the material. In particular, when a mixture contains only two isotopes it is possible to define the 'relative isotopic abundance' R by

$$R = N/(1-N)$$

which is equivalent to the ratio of the numbers of molecules of the two isotopes in the mixture. The effect of a separative element on this ratio is given by the 'single-stage separation factor', q, defined by

$$q = R_P/R_W \quad [2a]$$

For many processes this number is only slightly larger than one, so it is often more convenient to deal with the small quantity $g = q-1$, called the 'separation gain'. For example, suppose that a particular enrichment process is capable of extracting 51 per cent of the ^{235}U atoms present in the feed material while taking along only 49 per cent of the ^{238}U. For every 100 atoms of feed there are 100 N_F atoms of ^{235}U and $100(1-N_F)$ of ^{238}U, so the composition of the product will be 51 N_F ^{235}U and $49(1-N_F)$ ^{238}U. The relative isotopic abundance in the product will therefore be

$$R_P = \frac{51}{49} R_F$$

Similarly, the relative abundance of the tails is given by

$$R_W = \frac{49}{51} R_F$$

Then the single-stage separation factor is

$$q = R_P/R_W = \left(\frac{51}{49}\right)^2 = 1.0833$$

and the separation gain is

$$g = 0.0833$$

Entropy and separative work

The fundamental property which is being changed by the separation process is entropy. Entropy is a measure of the disorder of a system, and the product and tails from a separative element constitute a slightly less disordered system than the incoming feed material. In other words, the isotopes have been partially separated, so they are approaching the more orderly state of total separation. This means that the entropy of the material has been decreased by the separative process.

The second law of thermodynamics states that any decrease in the entropy of a system can only be accomplished with the expenditure of energy. Therefore, the separation element can be visualized as absorbing energy and converting it into order, or what is often called 'negative entropy'. The entropy change per unit of feed produced by a separative element such as the one in figure 5.1 is given by[1]

$$\Delta S = \tfrac{1}{2} K\, g^2 \theta(1-\theta)\, N_F(1-N_F) \tag{5.1}$$

where

$$\theta = P/F$$

is the ratio of product flow to feed flow (usually called the 'cut'). K is a constant whose value depends on the system of units for entropy. This expression for the entropy change is valid as long as the effect of a single separative element is small (i.e., as long as $g \ll 1$).

If the separation process were thermodynamically ideal (i.e., revers-

---

[1] This calculation assumes that the product and tails streams emerge at the same temperature and pressure as the feed stream.

ible) then a knowledge of the entropy change would allow one to calculate the rate of energy consumption of the separation element. The energy consumption is extremely important in comparing the costs of various enrichment techniques and in estimating the cost of producing reactor fuel or nuclear explosives. However, as will be seen in chapter 6, all known enrichment processes are far from ideal, and there is no simple connection between the energy requirements for a separating element and the entropy change it creates.

Another property of the entropy which makes it inconvenient for uranium enrichment is its dependence on the composition of the feed material. Equation 5.1 shows that the entropy change induced by a separative element is greatest when $N_F = 0.5$ and becomes very small when N_F is either close to zero or close to 1. So a given element will produce different entropy changes at different locations in a cascade, a property which is very inconvenient when one attempts to design a cascade containing thousands of elements.

The awkward properties of the entropy were recognized by the scientists who first formulated the theory of uranium enrichment in the early 1940s. As a substitute they invented the concept of 'separative work' or 'separative power' [3a] in order to have a quantity which could be attributed to a separative element independent of its position in a cascade and which was roughly proportional to the rate of energy consumption of the element. Instead of the entropy of a mixture, one defines the 'value' of a mixture by the formula

$$V(N) = (2N-1) \ln (N/1-N) \qquad (5.2)$$

where ln stands for the natural logarithm [2b]. This equation is plotted in figure 5.2. The rate of change in value produced by the separative element is called the 'separative power' (separative work per unit time) of the element and is given by

$$\Delta V = \frac{1}{2}F g^2 \theta (1-\theta) \qquad (5.3)$$

This quantity is measured in the same units as the feed flow, for example, kilograms of uranium per year, but it is normally expressed as a number of 'kilogram separative work units' (kg SWU) per unit time, for example, kg SWU/yr.

Suppose, for instance, that a separating element has a separation gain of 0.0833, a cut of one-half, and a capability to process 10 kg of uranium feed per hour. Such an element would be rated according to equation 5.3 at 76 kg SWU/yr. It is important to remember that for this concept to be useful the feed rate must be expressed in terms of the mass of uranium processed rather than the amounts of UF_6 or other compounds or mixtures.

The concept of separative work was invented in order to have a property of an element which was closely related to the rates of material flow and therefore to the energy consumption in a plant with many stages

Figure 5.2. The value function

Graph of equation 5.2 for the value function. Note that the horizontal scale is logarithmic in the ^{235}U fraction N. To compute $V(N)$ for $N>0.5$ replace N by $(1-N)$.

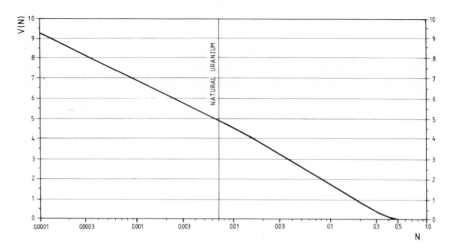

and small enrichment per stage. It was in fact invented for the analysis and optimization of gaseous diffusion plants. The smaller the number of stages and the greater the enrichment factor per stage, the less useful is the concept of separative work. For a single-stage process with an enrichment factor much greater than 1 the concept is essentially meaningless. There are other processes, such as chemical exchange, for which the concept of separative work is also not very relevant. In these processes the major energy inputs occur outside the cascade itself and therefore are unrelated to the separative power of individual elements. However, because separative work has a strong historical tradition in the field of uranium enrichment, it is still widely used as a unit of measurement of enrichment capacity for the purposes of comparing different processes. It will be used in this sense in this study.

III. A sample enrichment plant

Separative capacity

The concept of separative work can also be used to analyse a complete enrichment plant, treating it as a much larger black box (see figure 5.3(a)). As an example, one can consider a facility with a capacity of 1 000 t SWU/yr and ask what quantities of feed, product and tails are required.

The total separative work is computed by taking the difference between the total 'values' of the outputs and inputs. For example, the value of the product is calculated by multiplying the number of kilograms of product P by the value function evaluated for the particular enrichment achieved:

$$\text{Product value} = P \times V(N_P)$$

The value function can be obtained from equation 5.2 or the graph of figure 5.2. The total separative work done by the plant is

$$\Delta V = PV(N_P) + WV(N_W) - FV(N_F) \qquad (5.4)$$

Figure 5.3. Model enrichment facilities
  (a)   A 1000 t SWU/yr facility producing 3 per cent product and 0.2 per cent tails from natural feed.
  (b)   A 1000 t SWU/yr facility producing 90 per cent enriched product and 0.2 per cent tails from natural feed.

For both facilities feed, product and tails flows are shown in tonnes of uranium per year.

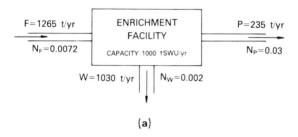

(a)

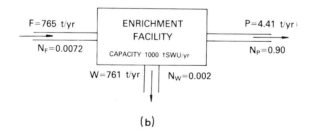

(b)

If it is assumed that no material is consumed or lost within the plant, then the total amounts of ^{235}U and ^{238}U which emerge must be the same as those which entered. This fact can be used to determine relations between the amounts of product and waste and the amount of feed. These relations are

$$F/P = (N_P - N_W)/(N_F - N_W)$$

$$W/P = (N_P - N_F)/(N_F - N_W) \qquad (5.5)$$

99

These relations can be inserted into equation 5.4 to give

$$\Delta V = P \left\{ V(N_P) + \frac{(N_P - N_F)}{(N_F - N_W)} V(N_W) - \frac{(N_P - N_W)}{(N_F - N_W)} V(N_F) \right\} \quad (5.6)$$

This formula is extremely useful and not as complicated as it looks. The trick is to be able to evaluate the value function, and this can be done with the aid of figure 5.2. The great power of the formula is that once the compositions of the feed, product and tails are specified, then all material flow rates are determined by the separative capacity. For example, suppose the facility of figure 5.3(a) is used to produce 3 per cent enriched reactor fuel from natural uranium, and that the tails assay is set at 0.2 per cent. These values of N_P, N_F and N_W can then be inserted into equation 5.6, and with the aid of figure 5.2 the quantity in brackets can be shown to equal 4.25. Therefore, it requires 4.25 kg SWU to produce each kilogram of product in this plant. With a total capacity of 1 000 t SWU/yr, the plant can produce 235 tonnes of reactor fuel—enough to supply the annual reload requirements of eight or nine large nuclear reactors[2].

This figure for product flow rate can then be inserted into equations 5.5 to determine the feed and tails flow rates of 1 265 and 1 030 t/yr, respectively. Note that the sum of the product and tails flows equals the feed flow. This necessary result provides a useful check on the calculations, the results of which are displayed in figure 5.3(a).

Commercial enrichment services

In all existing types of commercial enrichment processes the operating costs are dominated by factors directly related to the rate of production of separative work units. It has therefore become customary to think of the product of an enrichment plant in terms of SWUs rather than in terms of kilograms of product at some enrichment.

This is the principle underlying the concept of toll enrichment services. Under this scheme a customer contracts for a certain number of SWUs per year and then has the option of using them to make any combination of product and tails he desires within the limitations of the enrichment facility. As long as the plant is operating continuously at full capacity, thereby amortizing its capital costs at a maximum rate, the operator will be indifferent to the actual concentrations of product and tails. In fact it is common practice to operate large enrichment facilities in such a way that there are several inputs and outputs at various levels of enrichment. A typical commercial plant, therefore, is not confined to making only one product at only one tails assay. An interesting example of

---

[2] A 1 000 MW(e) pressurized water reactor requires about 27 tonnes of 3 per cent enriched fuel per year [4].

the creative use of toll enrichment services is given in chapter 8 (see Italy, p. 217).

Production of highly enriched uranium

In principle, the model enrichment facility of figure 5.3(a) could be adapted to produce 90 per cent enriched product. The results of this conversion are shown in figure 5.3(b) (p. 99). Notice that the quantity multiplying P in equation 5.6 has changed from 4.25 to 225.6, assuming that the tails assay has been kept at 0.2 per cent. This means that only 4.43 tonnes of product can be produced from 765 tonnes of feed, leaving 761 tonnes of tails. This gives a good idea of just how little ^{235}U is present in any quantity of natural uranium. But, however small this amount of product may seem relative to the feed and waste flows, it is not small in its importance. In fact, it can be shown from the data in table 1.1 (p. 5) that 4.4 tonnes of 90 per cent enriched uranium can be used to make between 250 and 300 nuclear weapons.

One other brief computation will serve to give a better idea of the quantities of material involved. A feed flow of 765 t/yr may seem like a large amount, but it must be pointed out that a year is a long time, and the actual material flows are not so large. One way to see this is to note that, for example, 765 t of uranium in the form of UF_6 at normal pressures and temperatures could be delivered to the plant through a single pipe less than 6 cm in diameter if the gas flow speed were 1 m/s. (It should be emphasized that this calculation was done only for illustration purposes. UF_6 is normally handled at somewhat higher temperatures and lower pressures, so actual pipes are not necessarily this small.)

In order to compare the production of low-enriched and highly enriched uranium, the enrichment facility has been treated simply as a black box with a certain separative work capacity. But actually to change a facility designed for the production of low-enriched uranium to one capable of producing highly enriched product requires entry into the black box to make certain changes. The above calculation assumes that these changes can be made instantaneously, with no loss of separative work. This is obviously an over-simplification, and different enrichment processes differ from this ideal in a wide variety of ways. This question of the ease of convertibility of a facility from low to high enrichment has obvious implications for nuclear weapon proliferation and is one of the key parameters by which different processes are assessed in chapter 2.

Setting the tails assay

Another important factor which can be studied with this model is the effect of changing the tails assay. This effect can be shown most dramatically by supposing that the production of 90 per cent enriched product was accomplished at 0.4 per cent tails instead of 0.2 per cent. In this case the multiplier of P in equation 5.6 would be reduced to 169.3, increasing the production of 90 per cent ^{235}U to 5.9 t. But the requirements for feed would also be greatly increased, from 765 to 1 654 t.

This example illustrates a general problem in the design of enrichment facilities, whether they are to be used for civilian or military purposes. The fixing of the tails assay is an economic compromise which balances the unit costs of feed material against the unit costs of separative work. If feed is cheap and SWUs expensive (e.g., if the cost of electricity is high) then a relatively high tails assay is advisable. But as the price of natural uranium rises and enrichment processes become more energy-efficient, the appropriate tails assay should drop to lower values (see also appendix 8B).

Another possible method for producing highly enriched uranium from this plant is to use low-enriched reactor-grade material as the feed. Using the above formulae, it can be shown that with $N_F = 0.03$, $N_W = 0.01$ and a total capacity of 1 000 t SWU/yr, an output of 19 tonnes of 90 per cent enriched product is achievable, over four times the amount obtainable from natural uranium feed. This suggests another way in which an ostensibly peaceful enrichment facility could be utilized for the production of weapon material. But again there are technical problems which make this more or less feasible for various processes. These are discussed in chapter 6, and their implications for proliferation are evaluated in chapter 2.

IV. Cascades

Stages

It is now time to go inside the enrichment facility and describe the way in which the many separating elements are arranged in a cascade. This task is difficult because each process has its own peculiar requirements for an optimum cascade, so these initial descriptions must be quite general and abstract. They will be made more concrete in the discussion of individual processes in chapter 6.

The basic concept of a cascade is illustrated in figure 5.4. which shows an array of 'elements' organized in 'stages'. The elements within a stage are said to be connected in 'parallel', that is, they all receive identical inputs and produce identical outputs which are fed into other stages. By arranging many elements in this way, large amounts of material can be processed in a given stage, even though an individual element may have a very small capacity. This suggests that the 'width' of the cascade (i.e., the number of elements in a given stage) is proportional to the total rate of flow of material passing through that stage.

Figure 5.4. A cascade

The arrangement of enrichment elements and stages in a cascade. Note the decreasing number of elements towards the product end of the cascade. Tails flows are not illustrated.

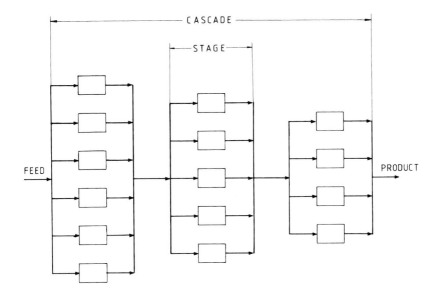

The various stages are connected to each other in series, that is, each stage receives input from the previous one and passes its output on to the next stage. Notice that the width of the stages is not constant, implying that the amount of material processed at each stage changes with the stage number. That this should happen is intuitively clear from the previous discussion, which showed that the rate of flow of product is always considerably less than the rate of flow of feed.

Figure 5.5 illustrates the way in which a cascade constantly recycles material to extract the ^{235}U. Each box in the figure now represents a stage, and the output of each stage consists of two streams — an enriched and a depleted stream, both of which differ in ^{235}U fraction from the input. The enriched stream is sent forward to provide part of the input for the next

Figure 5.5. A symmetric cascade [2c]

Product streams are represented by solid lines and tails streams by dashed lines. The number of enriching stages is $S + 1$ and stripping stages T.

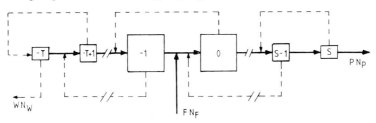

103

stage. Meanwhile the depleted stream is sent backward to serve as part of the input to a lower stage. So each stage is processing material of a given enrichment, part of which comes as enriched material from a lower stage and part of which comes as depleted material from a higher stage.

For most processes the enriched and depleted streams are each sent forward or backward by one stage; such a cascade is called 'symmetric' [2d]. In this case the 'cut' (see section II) will be slightly under one-half. But some processes (e.g., the aerodynamic nozzle process) use a different cut and asymmetrical amounts of enrichment and depletion in the two output streams. In this case these streams must be sent different numbers of stages forward or backward, and the cascade design becomes considerably more complex. This idea will be explored further in chapter 6, in the discussion of aerodynamic methods.

It can now be understood why a cascade must have two major sections—an enriching section above the point where the feed enters and a 'stripping' section below the feed point. The number of stages needed in each section is determined independently by the desired product and tails assays, and the widths of the stages in each section are determined by the flow rates of enriched and depleted materials. It is remarkable that this small number of parameters is sufficient to determine completely the shape of an 'ideal' cascade. This is a cascade that minimizes the ratio of separative work to product produced, by ensuring that streams of differing concentrations are never mixed together.

In an ideal cascade there is a simple relationship between the isotopic ratio at any stage and that of the previous and subsequent stages. If n denotes the stage number and R_n the relative isotopic abundance of material entering that stage, then

$$R_{n+1} = \alpha R_n$$
$$R_{n-1} = \frac{1}{\alpha} R_n$$

(5.7)

where α is called the 'enrichment factor' of a stage [2e]. It was equal to 51/49 in the example in section II.[3]

Equations 5.7 can be used to determine the isotope ratio at the $n+2$ stage since

$$R_{n+2} = \alpha R_{n+1} = \alpha^2 R_n$$

and in general the relationship is

$$R_{n \pm m} = \alpha^{\pm m} R_n$$

In particular if the feed is assumed to enter stage $n = 0$ and the product to

---

[3] Note that reference [2e] uses the symbol α^* for the quantity called α in this book.

emerge from stage S, then

$$R_P = \alpha^{S+1} R_F$$

Similarly, if the last stage of the stripping section is number T, it follows that

$$R_W = \alpha^{-T} R_F$$

Using these equations it is possible to compute the total number of stages (which equals $S+T+1$) required in an ideal cascade, once the product and tails assays are specified and the enrichment factor is known:

$$S+T+1 = \frac{\ln (R_P/R_W)}{\ln \alpha} \qquad (5.8)$$

To continue with the previous example, suppose $g = 0.0833$ and $\theta = 0.5$. Then $\alpha = 1.0408$ and $\ln \alpha = 0.0400$. If the cascade is to produce 3 per cent product with 0.2 per cent tails, then $R_P = 0.0309$ and $R_W = 0.002$. Equation 5.8 then shows that 68 stages would be required in such a plant. By solving for S and T independently, it can also be shown that $36(S + 1)$ of these stages must be in the enriching section and $32T$ in the stripping section.

The difficulties involved in producing highly enriched uranium can be partially illustrated by computing the number of stages needed to provide a product enriched to 90 per cent. In this case $R_P = 9.0$, and if the feed and tails assays are kept at their previous values, the above calculation can be repeated to show that 210 stages would be required in the full cascade, 178 of them in the enriching section.

Material flows

Another important property of a cascade is the amount of material which must be pumped through each stage in order to obtain a given rate of output. If L_n is taken to be the flow rate of material into the nth stage of an ideal cascade, then to a very good approximation

$$L_n = \frac{4P}{g} \frac{(N_P - N_n)}{N_n (1 - N_n)} \qquad (5.9)$$

in the enriching section and

$$L_n = \frac{4W}{g} \frac{(N_n - N_W)}{N_n (1 - N_n)} \qquad (5.10)$$

in the stripping section [2f].

Notice that the second of equations 5.5 in the previous section can be reproduced by equating the above two expressions for $n = 0$ ($N_0 = N_F$).

The quantity L_n can be computed for all values of n, leading to a profile of the cascade such as the one shown in figure 5.6. In this diagram, the vertical dimension is the stage number and the horizontal dimension is the total flow rate of material in each stage. Both cascades considered in this chapter have been drawn to show the dramatic contrast between their shapes. It should be noted, however, that the general shape of the cascade is similar in both cases and is in fact a characteristic shape for all ideal cascades. The shape has even been adopted as the corporate logo of Eurodif, the firm which owns the Tricastin gaseous diffusion facility (see figure 5.7).

Figure 5.6. Two ideal cascades

Scale models of two ideal cascades for the facilities of figure 5.3. The vertical scale is the number of stages, and the horizontal scale the stage flow rates in thousands of tonnes of uranium per year. The separation gain per stage is $g=0.0833$ and the stage enrichment factor $\alpha=1.041$. The stage cut θ is approximately one-half.

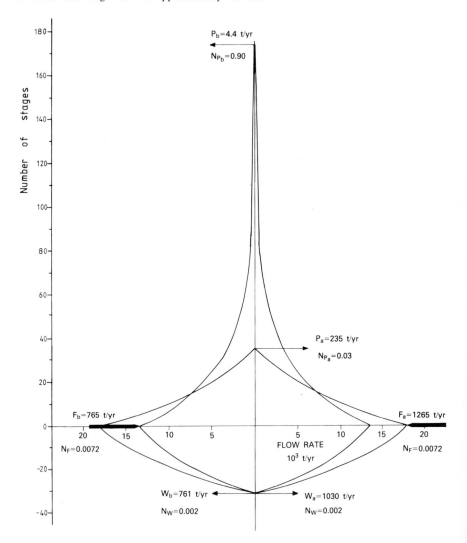

Figure 5.7. A cask of UF$_6$

A container of natural UF$_6$ feed material at the Tricastin gaseous diffusion plant. Note the Eurodif logo in the shape of an ideal cascade.

Source: *Nuclear Engineering International*, Vol. 25, No. 305, October 1980, p. 42; Eurodif S.A., Bagneux, France.

The most remarkable feature of the cascades in figure 5.6 is the enormous material flow required in the lower stages. To produce 235 tonnes of 3 per cent product per year (a flow rate of only 7.5 g/s) requires that about 36 000 t/yr (1.14 kg/s) be pumped through the lowest stage. Notice that this flow is 28 times as large as the feed flow, so that the latter makes only a very small contribution to the flow in the first stage. Most of the flow represents material recycled from the second stage and sent forward from the top stage of the stripping section. The flow differences are even more dramatic for 90 per cent enrichment, where the flow in the first stage is over 6 000 times as great as the product flow rate. It is the volume and energy required to carry and pump these enormous quantities of gas which make the gaseous diffusion and aerodynamic processes so large and expensive to operate.

The cascade shapes of figure 5.6 can be used to determine one more important property of the cascade. Since the vertical dimension of the graph measures the number of stages, and the horizontal dimension the flow per stage, then the area of the cascade profile must be equal to the total rate of material flow. This turns out to be directly proportional to the total separative power of the cascade. The relationship is simply

$$L_{\text{total}} = (8/g^2)\Delta V \qquad (5.11)$$

where ΔV is given by equation 5.6. For the 1 000 t SWU plant of figure 5.6 the total flow would be 1.15 million t/yr or 36.5 kg/s. This total flow rate can be seen from equation 5.11 to be very sensitive to the value of g. If g were to be decreased from the 0.0833 of this example to the 0.005 which is characteristic of gaseous diffusion, then the total flow would increase by a factor of almost 280 to 320 million t/yr in a 1 000 t SWU/yr plant. In such a plant the total material flow is more than a million times the production rate. On the other hand, a centrifuge plant with $g = 0.5$ would require a total flow of only 32 000 t/yr. This factor of 10 000 in flow reduction explains much of the attraction of the centrifuge.

One final interesting consequence of equation 5.11 is that the areas of the two cascade diagrams shown in figure 5.6 must be equal. In general any two ideal cascades which perform the same amount of separative work will have the same total flow, as long as they use elements with the same separation factor. So the rearrangement of a cascade from a low enriching to a high enriching configuration should not result in higher energy costs.

Inventory

The previous calculation has given the total flow rate and the flow rate per stage, but it has not specified how much material is actually present in any given stage or in the full cascade. The same flow rate could be produced by a large quantity of material pumped relatively slowly through a stage or a small quantity pumped very rapidly. In order to compute the material 'inventory' (or 'hold up') in steady-state operation, another characteristic of the process must be specified, the so-called 'transit time' or 'hold-up time' [2g]. This is the time it takes a given sample of material to pass through a single stage, and it is determined by the specific design of the separating elements.

Once the hold-up time t_h is specified, the total material inventory can be determined as the product of the total flow rate and the hold-up time. For example, if the total flow rate is 36.5 kg/s and the hold-up time is 10 s, then the inventory I is 365 kg of uranium. The formula for the inventory is

$$I = (8t_h/g^2)\Delta V \qquad (5.12)$$

Equilibrium time

Normally a cascade is started by filling all stages with material of the same initial isotopic composition. Then, with the product extraction valves closed, the pumps are started. As material is circulated through the multiple stages of the cascade the desired isotope begins to accumulate in the higher stages, but no product is extracted until the concentration of ^{235}U in the top stage reaches the desired level. During this period the cascade is said to be operating in 'total reflux', a term borrowed from the older technique of fractional distillation [5]. This concept is discussed in more detail below.

The enrichment at the output of the top stage is monitored, and when it reaches the desired level the output valve is opened slightly and a small flow of product is extracted. Over a period of time the product extraction rate is slowly increased in such a way that the product enrichment remains constant. Eventually, the cascade will reach its steady-state operating level, and product of the desired enrichment will flow out of the plant at the maximum rate consistent with its degree of enrichment.

The 'equilibrium time' of the cascade is conventionally defined as the time from the initial start-up to the point at which the product flow rate reaches half of its asymptotic value. This time gives a good estimate of how long it takes either to start up a new plant, to change enrichments in an operating plant, or to recycle enriched material through the plant for a second enrichment.

The equilibrium time t_e of an ideal cascade can be estimated from the formula [3b]

$$t_e = 8t_h/g^2 \ E(N_P, N_F) \qquad (5.13)$$

where E is a multiplying constant determined from the product and feed enrichments as follows [3b]:

$$E\ (N_P,\ N_F) = \frac{(N_P - 2N_PN_F + N_F)\ \ln\ (R_P/R_F)}{N_P - N_F} - 2 \qquad (5.14)$$

For the sample cascade considered here

$$E\ (0.03,\ 0.0072) = 0.34$$

and with $t_h = 10$ seconds and $g = 0.0833$, the equilibrium time is

$$t_e = 3\ 920 \text{ seconds}$$

or just over one hour. Since an ideal cascade gives the minimum possible equilibrium time for given values of product and feed enrichments, there is no way to reduce this number by rearranging the cascade elements [3c].

The inventory and equilibrium time are important parameters in evaluating the suitability of a facility for batch recycling to obtain high enrichments. The plant shown in figure 5.6 which can produce 3 per cent product from natural feed could, if the product were recycled, produce 11.7 per cent enriched material. With three more recyclings the product enrichment could be brought up to over 90 per cent.

But in order to recycle the output the plant must be shut down and all stages pumped out, refilled with the new feed, and then started up again. The minimum amount of material one can start with to fill the entire cascade is the 365 kg computed above. This represents about one month's production from the plant under the assumed operating conditions. But much more 3 per cent material than this must be produced in order to run the plant for the time required to make a sufficient amount of 11.7 per cent material.

If the implications of this schedule are followed through, it can be shown that batch recycling is a very inefficient method for producing highly enriched product. Batch recycling wastes large amounts of ^{235}U because the tails assays increase in every recycling after the first one. So, much more feed material and operating time are needed to produce a kilogram of highly enriched product than in a properly designed cascade. Whether such a waste of time and valuable resources is considered worthwhile depends, of course, on the strength of the motivation to produce highly enriched uranium.

Square and squared-off cascades

The ideal cascade considered up to this point is 'ideal' in two important ways: (*a*) it minimizes the ratio of total cascade flow to product flow, thereby producing the largest possible amount of product for a given enrichment, tails assay and separative capacity; and (*b*) it is the cascade with the shortest possible equilibrium time for a given product enrichment.

These are both highly desirable properties, but much more must be considered in the design of a commercial plant. In a practical facility the ideal is achieved when the cost of each SWU is minimum. Both total cascade flow and equilibrium time contribute to the cost per SWU. The former largely determines the power requirements for the cascade, while the latter affects the cascade's productivity and adaptability to variations in product requirements. It should also be noted that minimal cascade flows also imply minimal cascade inventory (cf., equations 5.11 and 5.12), and the amount of inventory has an effect on the capital costs of the plant.

But, the ideal cascade also has an important disadvantage, which can be seen by referring to figure 5.6 or equations 5.9 and 5.10. The variable shape of the cascade requires that every stage carry a slightly different flow from the ones adjacent to it. Therefore, no two stages can be the same size or use the same amount of power. To actually construct an ideal cascade of, say, 1 000 stages would require the construction of 1 000 different machines. Such a project is more akin to Gothic cathedral construction than to modern industrial design.

The solution to this problem in practical cascade design is to approximate the ideal cascade by a small number of 'square' cascade segments. A square cascade is one in which the flow rates are the same in all stages. It is easy to see that such a cascade would be represented by a rectangular shape on a graph with the same axes as figure 5.6. The advantage of a square cascade is obvious: all of the stages can be made identical, making possible substantial savings by standardizing their manufacture.

110

In principle it is possible to construct a square cascade which duplicates exactly the function of an ideal cascade. The product and tails assays and flows, and the total separative power, can be made the same. But in doing this with a square cascade the cut at each stage must be made the same as for all others, and when this is done it is no longer possible to avoid mixing process streams of different isotopic composition. This mixing results in losses of separative work, and the square cascade is therefore less efficient than an ideal cascade.

In a square cascade the ratio of product flow to interstage flow becomes an adjustable parameter which can be used to optimize the

Figure 5.8(a) and (b). Square cascades

The two ideal cascades of figure 5.6 are compared with square cascades of equivalent capacity, and product and tails assays. The square cascades have been optimized to the minimum number of stages per unit of product flow rate. Note that each square cascade has a larger area than its corresponding ideal cascade.

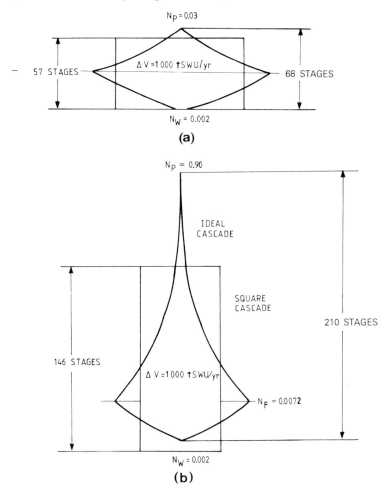

111

Figure 5.9. A squared-off cascade

An ideal cascade can be quite closely approximated by a small number of square cascades arranged as shown. This achieves a compromise between the low energy requirements and equilibrium time of an ideal cascade and the standardization of stage manufacture allowed by the square cascade.

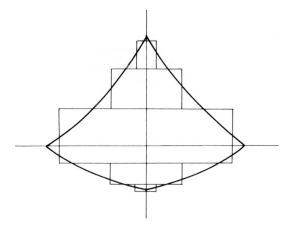

cascade performance [3d]. The procedure used is to specify the isotopic compositions of feed, product and tails and then to determine the ratio of product flow to interstage flow which minimizes the number of stages required for a given product flow. Since the energy consumption of the cascade is proportional to the number of stages, this procedure provides the most efficient possible square cascade within the input and output constraints.

The optimization procedure is mathematically complex, so only the results are presented here. Figures 5.8(a) and 5.8(b) show two square cascades designed for the same function as the two ideal cascades of figure 5.6. The latter are superimposed for comparison.

The results of this comparison are that a square cascade generally has fewer stages but a larger total flow than an ideal cascade with the same inputs and outputs. These differences become more pronounced as the product enrichment increases. In the 3 per cent case, the square cascade has 16 per cent fewer stages and a 28 per cent greater total flow, while in the 90 per cent case the square cascade has 30 per cent fewer stages but just over double the total flow rate of the ideal cascade. This means that just over half of the separative work performed by the stages is lost in remixing.

A major difference between ideal and square cascades is the equilibrium time. This is always larger for a square cascade, and for low enrichments can be estimated by multiplying the ideal equilibrium time by the ratio of product and feed assays [2g]. For example, the equilibrium time of the 3 per cent cascade would be increased by a factor of 4.2. For high enrichments the computation of equilibrium time becomes extremely complex, but an order of magnitude estimate can still be obtained by using the multiplying factor (N_P/N_F).

112

The ideal and the square cascade form the extremes of a range of cascade designs. In between are the squared-off cascades which use two, three or more square cascades of different widths to get better approximation to an ideal cascade. Figure 5.9 shows how such a cascade might be constructed.

The lengths and widths of the separate segments are determined by a complex procedure which is designed to take into account a wide range of capital and operating costs. In this way the savings in stage manufacturing costs can be balanced against the losses in separative efficiency, producing an optimum cascade.

Summary

The results of this section are summarized in table 5.1, which provides the information necessary to compute all of the important properties of a cascade for the four combinations of feed, product and tails assays given. All that is needed in order to estimate these quantities are the value of the

**Table 5.1. Cascade characteristics*

Cascade characteristic**	Basic parameter	Multiplication factor for input and output of compositions of			
		$N_F = 0.0072$ $N_W = 0.002$ $N_P = 0.03$	$N_F = 0.0072$ $N_W = 0.002$ $N_P = 0.50$	$N_F = 0.0072$ $N_W = 0.002$ $N_P = 0.90$	$N_F = 0.03$ $N_W = 0.01$ $N_P = 0.90$
Number of stages[a]	(l/g)	5.5–4.8[b]	12.4–9.6	16.8–12.2	13.6–10.2
Equilibrium time[c]	(t_h/g^2)	2.7–11.6	24–1 690	41–5 200	30–900
Feed flow	(P)	5.4	95.8	173	44.5
Tails flow	(P)	4.4	94.8	172	43.5
Separative power	(P)	4.25	121	226	52.2
Stage flow[d]	(P/g)	12.8–9.2	276–176	500–306	120–74
Total flow	(P/g^2)	34.0–44.0	971–1 670	1 805–3 720	418–744
Inventory[e]	(Pt_h/g^2)	34.0–44.0	971–1 670	1 805–3 720	418–744

* Applies only to cascades in which $g \ll 1$.
** Each cascade characteristic is completed by multiplying the 'basic parameter' in column 2 by the appropriate multiplication factor from columns 3 to 6 (see examples in text).
[a] Includes enrichment and stripping sections.
[b] The first number in each column refers to an ideal cascade and the second to a single, optimized square cascade. All squared-off cascades give values somewhere between these two extremes.
[c] Computed for enriching section only.
[d] Value for ideal cascade in the maximum stage flow at stage number 0 (see figure 5.5).
[e] Inventory in kilograms of uranium. To obtain inventory of ^{235}U, multiply by average fraction ^{235}U content of material in cascade.

separation gain g and the hold-up time t_h for the process in question. These values are collected in table 6.3 (p. 188).

A few points must be emphasized regarding the use of table 5.1. First, note that the two properties above the double line depend only on the stage characteristics and not on the capacity of the plant. An interesting consequence of this is that the equilibrium time of a cascade is independent of the amount of product to be produced. The quantities below the double line are all proportional to the production rate and must be multiplied by P to obtain the values for the full cascade.

A second point is that table 5.1 is only useful for cascades in which g is small compared to 1[4]. It works well for gaseous diffusion, aerodynamic and chemical-exchange techniques and is a useful approximation for the centrifuge. But it is not useful for the electromagnetic or laser processes, which produce very large single-stage enrichments. These latter processes must be analysed separately using the equations developed in section II. Generally with such processes the only useful property is the number of stages needed for a given enrichment. Concepts such as separative power, equilibrium time and inventory are not particularly useful for these processes.

Finally, all values in the table are based on the assumption that each stage cut is very close to one-half. When this is not true, as in the case of aerodynamic processes, an extra step must be performed (see chapter 6, section IV).

As an example of the application of table 5.1, consider a 1 000 t SWU/yr enrichment facility used to produce 90 per cent enriched product using 3 per cent feed and 1 per cent tails assay. Suppose this facility uses the hypothetical technology considered in this chapter with $g = 0.0833$ and $t_h = 10$ s. Using the multiplication factors from the last column of table 5.1 it can be shown that the number of stages needed would range from 163 for an ideal cascade to 122 for a square cascade. The equilibrium time would be anywhere from 43 200 s (12 hours) to 1.3×10^6 s (15 days). This wide range of equilibrium times is characteristic of high enrichments, and it provides a considerable incentive for using a squared-off cascade to approach more closely the minimum value.

In this facility 52.2 kg SWU would be needed to make each kilogram of product, so the output of the plant would be 19.2 t/yr. This implies a feed stream of 854 t/yr and a tails stream of 835 t/yr. None of these quantities change in going from an ideal to a square cascade.

The maximum stage flow in the ideal cascade would be 27 700 t/yr (0.88 kg/s), and the constant stage flow in the square cascade would be 17 060 t/yr (0.54 kg/s). The total cascade flow would be either 1.16×10^6 t/yr (36.7 kg/s) for the ideal case or 2.06×10^6 t/yr (65.1 kg/s) for the square case. For a process in which energy consumption is proportional to

---

[4] In particular, for ideal cascades the approximation $\ln \alpha = g/2$ has been made.

total flow, the square cascade would use almost 80 per cent more energy than the ideal. Here is another incentive for the use of a squared-off cascade. Finally, the inventory of uranium in the facility would be either 367 kg (36.7 kg/s × 10 s) or 651 kg.

Table 5.1 is intended for use in connection with table 6.3. In the latter table are collected the relevant parameters for all of the processes described in chapter 6. By using these two tables the reader can construct a model facility of any size using any desired process. The four combinations of N_F, N_W and N_P in table 5.1 were chosen to correspond to enrichment levels which are relevant to commercial or military tasks. The one exception is the second column, in which an enrichment of 50 per cent was chosen simply to give a rough idea of the rate of change of various parameters as final product enrichment is varied. If the reader wishes to examine other combinations of enrichments, all the formulae necessary for an ideal cascade are already provided. The only thing missing is the square cascade optimization, for which the reader is referred to the works of Brigoli [2a] or Cohen [3].

V. Reflux

As was noted above, the ratio of product flow to stage flow in a square cascade is an adjustable quantity. The smaller the rate of product extraction the higher the achievable enrichments for a given number of stages, and, conversely, higher production rates imply lower enrichments. These parameters can be varied in practice over a rather wide range.

The term 'reflux' refers to that portion of the stage flow at the top of a stage or cascade which is sent back down the stage or cascade to be reprocessed. The mechanism for accomplishing reflux is different for each technique but in every case the major requirement is that the reflux takes place with as little loss of uranium as possible. The importance of this can be seen in the model facility of section IV where in the square cascade the interstage flow was 900 times the product flow. That is, the 'reflux ratio' is 900 to 1.

Now suppose that for some reason 0.1 per cent of the uranium is lost in the reflux process. This apparently small loss turns out to be approximately equal to the product flow itself. This represents a reduction in plant output of 50 per cent as a result of a 0.1 per cent reflux loss. Of course this effect is greatly amplified at the high enrichments considered in this example, but even in commercial plants, especially those with very low enrichments per stage, this is a very important factor in making an enrichment process economically competitive. As each method is discussed in chapter 6, attention will be called to its reflux mechanism.

Another important use of reflux is in changing the production rate and isotope concentration of the product by adjusting the reflux ratio. If the rate of product extraction is decreased in a cascade, and the cascade continues to run at its rated capacity, the enrichment of the product will increase. Normally a cascade is optimized to produce maximum separative power for a given product and tails assays, so any variation from these values results in a somewhat less efficient cascade. In practice, however, these sacrifices are not great as long as variations from design assays are not too large, and cascades are often adjusted to produce product and tails of varying compositions [2h].

However, if a cascade operator is willing to accept more serious losses in efficiency and a low production rate, the product assay can be increased substantially. An estimate of the range of flexibility can be made by computing the enrichment which can be achieved under conditions of total reflux, that is, when no product is removed. To compute this, one first calculates the overall design enrichment factor of the plant

$$\alpha = R_P / R_F$$

Cascade theory shows that the enrichment factor under total reflux conditions is the square of this [2i], so the maximum value of R which can be obtained under total reflux is

$$R_{Pmax} = \alpha^2 R_F \qquad (5.15)$$

As an example consider an ideal cascade used for producing 3 per cent product from 0.72 per cent feed:

$$\alpha = 0.031/0.00725 = 4.28$$

and

$$\alpha^2 = 18.3$$

So the maximum R_P attainable is

$$R_P = 18.3 \times 0.00725 = 0.133$$

corresponding to a product enrichment of 11.7 per cent ($N_P = 0.117$). This example illustrates the fact that quite dramatic increases in product enrichment are possible in commercial cascades.

VI. Properties of uranium metal and UF_6

"Uranium is a heavy, silvery-white metal, which is pyrophoric (ignites spontaneously) when finely divided" [1a]. The same high degree of chemical reactivity which causes powdered uranium to burn spontaneously in air makes it highly corrosive to most materials when in the form of a metallic vapour. It is in this form that it is used in the plasma and atomic

116

vapour laser separation processes.

To vaporize metallic uranium requires substantial quantities of energy. The melting point of pure uranium metal is 1 132°C and the boiling point of liquid uranium is 3 818°C [1b]. Because the boiling point is so high the rate of evaporation of liquid uranium remains very low even at very high temperatures, but as the temperature is increased to speed up evaporation, the chemical reactivity of the liquid and vapour are made even stronger. For these reasons the production and handling of uranium vapour pose severe technological problems.

The uranium atom has six electrons outside its highest filled orbital shell. These are the electrons which are most often involved in both the

Figure 5.10. A sample of UF$_6$
The lucite cylinder encloses a vial containing about 32 grams of solid UF$_6$.

Source: Oak Ridge National Lab., Oak Ridge, Tennessee, USA.

absorption and emission of light and chemical reactions. Because of the enormous number of possible combinations of electrons and energy levels, the optical spectrum of uranium is extremely complex. In 1976 spectroscopists had already identified over 900 energy levels, 9 000 optical transitions and as many as 300 000 visible spectral lines [6]. Each of these lines is a potential candidate for excitation by a laser.

The six outer electrons also contribute to a rich and complex chemistry for uranium, which can form compounds in any of four oxidation states — U(III), U(IV), U(V) and U(VI) [7]. In each case the Roman numeral gives the number of electrons which the uranium atom contributes to covalent chemical bonds. The most commonly used forms of uranium are U(IV) and U(VI). Compounds of these forms are used in all of the major chemical-exchange techniques and, of course, in all of the processes that use UF_6. In addition, the material used to provide uranium ions for the calutron is uranium tetrachloride (UCl_4) [8]. The one exception to this rule seems to be the formation of uranium pentafluoride (UF_5) as a product of the molecular laser separation process.

The most important compound of uranium for the enrichment process remains uranium hexafluoride (UF_6). This is a colourless solid at room temperature (see figure 5.10). At atmospheric pressure it 'sublimes', that is, changes directly from the solid to the gaseous form at 56.5°C [2j]. In this way UF_6 behaves very similarly to solid carbon dioxide ('dry ice') but, of course, with a considerably higher sublimation temperature.

The molecular weight of UF_6 is either 349 or 352 atomic units, depending on whether it contains ^{235}U or ^{238}U. There are no naturally occurring isotopes of fluorine to obscure the direct connection between the isotopic and molecular masses. The structure of the UF_6 molecule is highly symmetrical, with the fluorines arranged around the central uranium atom along three mutually perpendicular axes (figure 5.11). This gives the molecule a complex but theoretically well understood spectrum of vibrational excitations [9]. Some of these vibrational transitions can be excited by lasers, and the difference in vibrational energies between the two isotopic forms of UF_6 allows this excitation to be made specific to $^{235}UF_6$.

Chemically, UF_6 is a highly reactive substance. It is a strong fluorinating agent and, for example, reacts violently with water and many organic compounds, such as oils and lubricants. For this reason all systems used for carrying or processing UF_6 must be extremely clean and free of leaks. One result of the search for non-reactive lubricants for UF_6 compressor seals and bearings has been the development of fluorocarbon and chlorocarbon materials such as 'teflon' (polytetrafluoroethylene) [2k].

UF_6 is highly corrosive to many metals, and generally only nickel or aluminium or their alloys are suitable for UF_6 handling [2m]. Hydrogen is another material that reacts with UF_6, but this reaction is slow at room temperatures and therefore does not appear to interfere with the use of a $UF_6 - H_2$ gas mixture in aerodynamic separation methods [10]. However, it

118

Figure 5.11. A UF$_6$ molecule

The uranium atom is in the centre and the six fluorines are arranged symmetrically on three perpendicular axes. This high degree of symmetry gives the UF$_6$ molecule its characteristic infra-red vibrational spectrum.

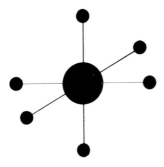

does seem that great care would have to be taken to prevent leakage of this mixture and possible explosions caused by sparks or flames.

In summary, neither uranium vapour nor UF$_6$ are particularly easy substances to work with on an industrial scale. But in neither case are the problems insurmountable, and after almost 40 years of large-scale treatment of UF$_6$ this particular art has few remaining mysteries.

Chapter 6. Enrichment processes

I. Introduction

This chapter presents brief, quantitative descriptions of a number of uranium enrichment processes that have been developed or are undergoing serious research and development. There exist, in fact, literally hundreds of ideas for separating uranium isotopes, and those of other elements as well. New ideas are still being proposed at a steady rate, and patents for new methods are granted regularly in many countries.

However, only 10 of these techniques have shown sufficient promise to have either been developed for commercial or military use or to be the subjects of intense research and development efforts. It is the purpose of this chapter to describe these enrichment methods in enough detail to allow an assessment of the potential contribution each process might make to further nuclear weapon proliferation.

Most of the technical details of the processes are still secret, so it is necessary for the descriptions that follow to focus on those features which can be inferred from a basic physical understanding of the process. Such analyses can occasionally be misleading, but in the great majority of cases it is possible to derive useful information, especially for the purposes of policy analysis, without detailed engineering data. Since the policy problem is an important one, there seems to be no alternative to doing the best that can be done with the data which are available.

II. Gaseous diffusion

Basic principles

The basic physical principle underlying the gaseous diffusion method is the so-called 'equipartition principle' of statistical mechanics. This principle states that in a gas consisting of several types of molecules each type will have the same average energy of motion (kinetic energy). This equality of average energies is attained and preserved by the enormous number of collisions between molecules which are taking place at all times in the gas. These collisions ensure that any excess energy which may have been associated with one component will rapidly be shared equally with all the others. This equal sharing is called thermal equilibrium. The kinetic energy KE of a molecule of mass m is related to its velocity v by the formula

$$KE = \tfrac{1}{2} mv^2$$

Therefore, molecules which have the same average kinetic energy will have average velocities which differ in inverse proportion to the square roots of their masses. Using the symbol $\langle \rangle$ to denote the average, the relationship can be written

$$\langle KE_1 \rangle = \langle KE_2 \rangle$$

$$\tfrac{1}{2}m_1 \langle v^2_1 \rangle = \tfrac{1}{2}m_2 \langle v^2_2 \rangle$$

or

$$\langle v_1 \rangle / \langle v_2 \rangle = \sqrt{m_2/m_1}$$

For uranium hexafluoride gas made up of $^{235}UF_6$ and $^{238}UF_6$ the respective molecular masses are 349 and 352, so the ratio of the velocities is 1.0043, with the lighter $^{235}UF_6$ molecules moving very slightly faster *on the average*. This last phrase is important, because the molecules move with a wide range of velocities, and the equipartition principle applies only to averages over large numbers of molecules.

The gaseous diffusion method of separation exploits this slight difference in average velocities by forcing the gas mixture to diffuse through a porous barrier under a pressure difference. The barrier is a thin wall of solid material containing many very small holes or passageways. The faster molecules will encounter the holes more often than the slower ones and will therefore be slightly more likely to pass through, causing the gas which emerges on the other side of the barrier to be enriched in the lighter isotope. This method is one of the oldest for separating isotopes and was first used by Aston in 1920 to partially separate isotopes of neon [11a].

The number of molecules of each type which emerge on the product side of the barrier is proportional to the rate of flow through the barrier of that type of molecule. This in turn depends on both the number of each type present on the feed side and their average velocities. So the ratio of numbers emerging on the product side is

$$R_P = N_{235} \langle v_{235} \rangle / N_{238} \langle v_{238} \rangle = R_F \sqrt{(M_{238}/M_{235})} \qquad (6.1)$$

Therefore, the ratio of the average velocities is the ideal enrichment factor α, defined in chapter 5 (see equation 5.7, p. 104). The actual enrichment factor is less than this for two reasons. First, the concentration of the desired isotope on the feed side is not constant; it gradually decreases as the enriched mixture diffuses through the barrier. If the cut is assumed to be one-half, then this effect reduces the ideal enrichment gain $\alpha - 1$ by a factor of 0.69. An ideal cascade containing ideal separation elements would therefore have $\alpha = 1.00297$, corresponding to a stage separation factor $q = 1.00595$ (see chapter 5). This is further reduced by the separation 'efficiency', to be discussed below.

The diffusion barrier

The central problem in the use of gaseous diffusion is the manufacture of a suitable barrier material. The difficulty in making the barrier can be appreciated if one lists the properties it must possess.

1. The average diameter of the holes (pores) in the barrier must be much less than the average distance travelled between collisions (the 'mean free path') of a molecule. If this is not satisfied then a molecule is likely to suffer one or more collisions near the entrance to the pore and inside the diffusion channel. This would cause exchanges of energy with other molecules and tend to cancel out the slight velocity difference between light and heavy molecules. At the same time the holes must be large enough to allow the gas to pass through at a reasonable rate.

2. The barrier must be very thin so as to have an adequate permeability at reasonable pressure, but it cannot be so thin as to break under the necessary pressure difference across it. It is desirable to have this pressure difference as large as possible to increase the rate of flow through the barrier.

3. The barrier material must be highly resistant to corrosion by the very corrosive gas UF_6. Any corrosion which occurs will cause plugging of the tiny holes in the barrier.

Requirements 1 and 2 can be made more quantitative by considering the actual properties of UF_6 gas. A UF_6 molecule has a diameter of about 0.7 nanometres (nm), and at a pressure of about one-half atmosphere and a temperature of 80°C the average separation between molecules is about 5 nm. The mean free path under these conditions is about 85 nm, so the average pore opening must be somewhat less than this, say about 25 nm. Using a rough geometrical argument it can be shown that such pore dimensions can be obtained by packing together spheres whose diameters are about 100 nm (0.1 μm). Figure 6.1 shows the relationship between these quantities drawn roughly to scale. Figure 6.2 shows an actual

Figure 6.1. Model diffusion barrier

This schematic diagram shows a barrier made of closely packed spheres 100 nm in diameter, giving average pore openings comparable to circles 25 nm in diameter. The average spacing between UF_6 molecules (the small dots below the barrier) is about 5 nm, and the mean free path of the UF_6 molecules is 85 nm.

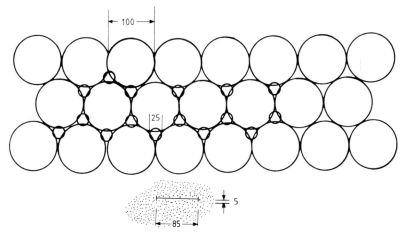

Figure 6.2. A sintered nickel barrier

Photomicrograph of a piece of sintered nickel barrier cut parallel to its surface.

Source: Commissariat à l'Énergie Atomique, Paris, France.

photomicrograph of a piece of barrier made from nickel powder which has been sintered (i.e., packed together under high pressure and heated to a high temperature).

It is now easy to understand why a barrier is quite difficult to produce. The actual methods used by various countries are classified, but it is known that the United States uses sintered nickel powders [2n], while those in the new French Tricastin plant are "ceramic" [12].

Japanese researchers have experimented with nickel, aluminium and Teflon barriers [13a] and although the details of barrier manufacture are generally held to be secret, some publicly available Japanese patents are quite explicit in their descriptions of proposed methods [14].

Whatever the material, it must be bonded under high pressure and temperature into sheets only a few microns thick. These very thin sheets must be able to withstand pressure differentials of the order of 0.3 to 0.5 kg/cm^2 for many years without failure. Generally this requires a carefully designed supporting structure, and possibly even multilayer barriers composed of materials with different porosities and strengths [2o]. That this kind of reliability is indeed achievable can be seen in the record of US plants in which "barrier failures are too rare to justify maintaining separate records" [15a].

The barrier must be assembled in a way which will maximize its area of contact with the gas. In US gaseous diffusion stages this is done by manufacturing the barrier in the form of sheets of small tubes assembled in cylindrical tube bundles [2n].

The performance of a barrier depends not only on its own properties but also on the pressure and temperature of the gas which comes into contact with it. For example, as the pressure on the feed side increases, the mean free path gets smaller and the probability of collisions in or near the pores is increased. This reduces the efficiency of the barrier. It is also desirable to keep the pressure on the product side (back pressure) low in order to prevent too many molecules from diffusing backwards through the barrier. High temperatures would make the molecules move faster, thereby increasing the diffusion rate, but raising the temperature also forces the compressors to do more work for a given amount of gas. So the temperature must be kept as low as possible without allowing the gas to condense or solidify. Keeping the temperature low also reduces corrosion problems.

A gaseous diffusion stage

In the design of a separation stage all of these factors must be balanced in an optimum way. The operating pressures and temperatures, the properties of UF$_6$ gas, and the structure of the barrier all combine to produce a barrier efficiency e_B, which is a number somewhat less than 1. The actual separation gain is related to the ideal value by

$$g = e_B g_{ideal}$$

The structure of the barrier also determines its 'permeability', or the rate of flow of gas through a given barrier area for a given pressure difference across it. Using data obtained from experiments with argon it is possible to estimate the dependence of permeability on pore size [13b]. Knowing the permeability of the barrier and its efficiency, it is then possible to compute how much separative power can be produced by a given area of barrier material. As an illustration assume that the barrier efficiency is about 0.7. Then the actual separation gain will be

$$g = 0.7 \times 0.00595 = 0.0042$$

which can be shown to lead to about 1 kg SWU per square metre of barrier per year. Therefore a plant such as the one at Tricastin, with a total capacity of 10.8 million SWU/yr, might have a total barrier area of 10^7 m^2, or about 10 square kilometres. A single large stage with a capacity of about 12 000 SWU/yr will have a barrier area of roughly 1.2 hectares (3 acres).

The individual tubes which make up the barrier must be small enough to provide a large surface area for diffusion but large enough to permit easy flow of the process gas. Again, no information is available on the size of the tubes, but if it is assumed that the tubes are about 2 m long and 1 cm in diameter, then about 160 000 of them would be used in such a stage. This can be compared with some of the early US stages which contained several thousand tubes each [16].

Using the above value of g in table 5.1 (p. 113), it can be shown that an ideal gaseous diffusion cascade for the production of 3 per cent enriched product with a 0.2 per cent tails assay would have about 1 290 stages.

Table 5.1 can also be used to calculate the total cascade flow and the flow per stage. For a 10^7 SWU/yr gaseous diffusion plant the product flow P would be 2.4×10^6 kg/yr, and the total cascade flow would be between 4.6×10^{12} and 6.0×10^{12} kg/yr. If the cascade were squared off, a typical large unit might carry a flow of 5.5×10^9 kg/yr or 174 kg U/s. This implies a flow through a given stage of 250 kg UF$_6$/s, a value comparable to the rated capacities of the largest compressors at Tricastin (190 kg/s) [17a].

It should be emphasized once more that the numbers used here are only estimates designed to produce approximate values for stage and plant parameters. The precise values of numbers such as barrier efficiency are well-guarded secrets.

In a typical gaseous diffusion stage (see figure 6.3) feed enters the diffuser at a pressure of between one-third and one-half atmosphere [18a], and half of it is allowed to diffuse through the barrier tubes. The low pressure diffused gas is then cooled by passing it through a heat exchanger and drawn into a compressor where it is compressed and mixed with the depleted material from a higher stage. The precooling of the gas is necessary because the compression heats it substantially, and most of the compression energy must be rejected to maintain a steady operating temperature. This wasteful procedure accounts for the high energy

Figure 6.3. Gaseous diffusion stages

Three of the gaseous diffusion stages used at the French Tricastin facility shown connected in a symmetric cascade. In the stage at the right the three labelled components are: 1. diffuser, 2. heat exchanger, 3. compressor. Note that the product gas from the diffuser is cooled in the heat exchanger before it is compressed and sent to the next stage. This explains why the product and tails flows do not match those of figure 5.5.

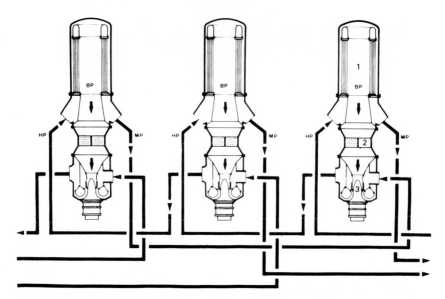

Source: Petit, J.F. (Eurodif), in *Nuclear Power and its Fuel Cycle*, Proceedings of an International Conference, Salzburg, 2–13 May 1977 (IAEA, Vienna, 1977), p. 114.

consumption of gaseous diffusion plants, which is generally between 2 300 and 3 000 kilowatt hours for every SWU produced [2p].

The output of the compressor is sent forward to the next stage while the depleted half of the gas is drawn off through a separate port and sent back two stages where it is mixed with the diffuser output and recycled.

Modern compressors are of the axial flow type, similar to those used in jet aircraft engines (see figure 6.4). The compressor rotor rotates at high speeds and must handle large flows of corrosive UF_6 at relatively high temperatures for long periods (many years) of continuous operation. And because the entire UF_6 carrying system must be leak-proof, the compressor shaft bearings require special rotating seals and the use of nitrogen gas to isolate the UF_6 from the shaft lubricants [2q]. The giant compressors used at Tricastin have all been nickel-plated to prevent corrosion [17b].

The barrier and the compressors are the key components in a gaseous diffusion plant. All the rest is really just some high-quality plumbing. Detailed design and manufacturing data on the barrier and compressors are classified, but history has shown that all countries which have set out to build a gaseous diffusion plant have succeeded. This suggests that the secrets are probably less important in inhibiting the further spread of

Figure 6.4. Rotor of an axial flow compressor

Source: Eurodif S.A., Bagneux, France.

gaseous diffusion technology than the enormous capital costs and industrial effort involved.

It remains to provide an estimate of the hold-up time in a gaseous diffusion stage. One way to do this is to use an estimated value for the 'specific hold-up', which is defined as the amount of uranium which must be kept in inventory to provide a separative power of 1 SWU/yr. One value which has been given is a specific hold-up of "not higher than 0.1 kg U/kg SWU/yr" [2r]. This can be used along with the value of $g = 0.0042$ in table 5.1 to show that $t_h = 6$ seconds. Given the uncertainty in this calculation it would be prudent to use $5-10$ seconds as a range of possible values. Using this range the equilibrium time of a commercial cascade making 3 per cent product comes out to a minimum of 11 days if the lower value of t_h is applied to an ideal cascade. A more reasonable estimate for a real cascade might be roughly three weeks.

The low single-stage enrichment factor and relatively long hold-up time combine to make gaseous diffusion a very large-scale, capital- and

127

energy-intensive process. To produce nuclear weapon-grade uranium from natural feed would require over 3 500 stages and an equilibrium time of at least a year. Production of the barrier, compressors, and piping would be major industrial undertakings, and the construction of the facility would put a heavy drain on the resources of most countries of the world. So, even though gaseous diffusion has been the major historical contributor to proliferation, it seems unlikely to continue to play this role in the future, especially in view of the capabilities of some of the other enrichment techniques under development or already in use.

III. The gas centrifuge

Basic principles

A good model for understanding the way in which a centrifuge separates a mixture of isotopes is to imagine a sample of gas in a room under the influence of gravity. Since each molecule is being pulled downwards, a certain amount of work must be done to lift the molecule to some height h. This work represents an increase in the molecule's 'potential' energy, and this change is given simply by

$$PE = mgh$$

where m is the molecule's mass and g is the acceleration due to gravity, equal to 9.8 m/s^2 at the Earth's surface. If the temperature in the room is the same everywhere, molecules at all heights have the same average kinetic energy, but those near the ceiling have a higher potential energy than those near the floor. The same theory that predicts the equipartition of energy (see chapter 1) predicts that higher energies are less probable than lower ones. It predicts that the density of particles near the ceiling will be less than the density near the floor by the factor

$$N(h)/N(0) = \exp -(mgh/RT) \qquad (6.2)$$

where R is called the gas constant (8.3 joules per degree Kelvin), and T is the temperature measured from absolute zero (0 K = $-273°$C). If two different species are present in the mixture, then an equation like 6.2 can be written for each one and the ratio taken:

$$[N_1(h)/N_1(0)]/[N_2(h)/N_2(0)] = \exp - [(m_1 - m_2)gh/RT] \qquad (6.3)$$

If the factors on the left are slightly rearranged, the ratio can be seen to be equivalent to the ratio of the relative isotopic abundances at the heights h and zero. So this equation can be rewritten as

$$R(h)/R(0) = \exp - [(m_1 - m_2)gh/RT] \qquad (6.4)$$

In this form it is seen to be nearly analogous to equation 6.1 for gaseous diffusion.[1] Note that if subscript 1 refers to the lighter species, then equation 6.4 is consistent with the relative abundance of this species increasing with h. This calculation suggests that it should be possible in principle to separate isotopes of uranium by filling a room with UF_6, allowing it to come to thermal equilibrium, and then simply skimming off the top portion of the gas. Experiments similar to this were in fact performed over 80 years ago [2s]. However, this is not a practical process for uranium, since the effect is extremely small. For example, in a room 3 metres high at normal temperature the ideal separation gain for UF_6 would be only 3.4×10^{-5}, over 100 times smaller than for gaseous diffusion.

It was recognized very early that a rapidly rotating centrifuge could provide a much stronger force field and therefore increase the separation gain many times. In a centrifuge the acceleration of gravity is replaced by the 'centrifugal' acceleration, and the equation corresponding to 6.4 can be shown to be

$$R(r)/R(0) = \exp\left[(m_1 - m_2)(\omega r^2)/2RT\right] \qquad (6.5)$$

where r is the distance from the centre of the centrifuge, and ω is the angular velocity in radians per second. Notice that the sign of the exponential factor has changed, implying that the isotopic abundance of the lighter species increases towards the centre of the centrifuge. The wall of the centrifuge is then analogous to the floor in the previous example.

To see what equation 6.5 predicts for uranium enrichment consider a hypothetical centrifuge with a radius of 10 cm and an angular frequency of 800 revolutions per second ($\omega = 800 \times 2\pi = 5\,000$ rad/s). The acceleration at the wall of this centrifuge is 2.5×10^6 m/s^2 or more than 250 000 times as strong as gravity. Using $m_1 = 0.349$ kg and $m_2 = 0.352$ kg and assuming a temperature of 330 K (57°C) the ratio of abundance between the centre and outer wall is found to be 1.147. The ideal value of $g = 0.147$ for this centrifuge can be seen to be over 16 times as large as the corresponding value for gaseous diffusion. Note also that the separation gain depends on the simple difference between the isotope masses, not on their ratio. This means that the advantage of the centrifuge over gaseous diffusion improves as the isotopic masses increase.

Modern centrifuges

All of these advantages were recognized by the early workers in isotope separation, and centrifuge research was actively pursued in initial efforts in both the USA and Germany during World War II (see p. 16). But very

---

[1] Actually the exponential quantity in equation 6.4 must be interpreted as a separation factor rather than an enrichment factor (see p. 95). Therefore equation 6.4 is analogous to the square of equation 6.1.

quickly a number of disadvantages were discovered, and it has taken many years of active research and development, and a number of important technological advances, to bring centrifuges into competition with gaseous diffusion. The following is a list of these problems and their solutions [19a].

1. The ideal separation factor derived above is actually the ratio of relative isotopic abundances at the centre of the centrifuge to that at the wall. To take full advantage of this difference, product would have to be extracted at the centre and tails at the wall; but the rapid rotation of the centrifuge causes virtually all the gas to concentrate near the wall. For example, in the above centrifuge the pressure of the gas at the wall can be shown to be 40 million times that at the centre [20a]. So it would be useless to attempt to extract any product, no matter how enriched, from near the centre. In fact all the separation must take place in a narrow annular region near the wall of the centrifuge, and this greatly reduces the separation effect unless a 'countercurrent' flow pattern is created.

The countercurrent flow is a form of internal reflux which causes a continual recirculation of the gas in the centrifuge, allowing a long-term exchange to take place between layers of differing isotopic concentrations [21] (see figure 6.5). The countercurrent flow is induced by aerodynamic interactions between the rotating gas and the bottom scoop. The lighter component tends to concentrate in the inner, rising layer and therefore at the top, while the heavier component concentrates at the outside and bottom. Feed is injected near the centre.

2. Normally when an object is rotated at high speeds it must be carefully balanced to prevent wobbling and vibration. At the rotational speeds necessary to get useful separation factors for UF_6 this problem becomes very severe. In addition to these difficulties there is a problem of 'critical' rotation frequencies at which the centrifuge can be set into a kind of resonant vibration which can grow to large amplitudes and destroy the centrifuge. Since it is desirable to operate the centrifuge above these critical frequencies if possible, the problem is to find ways both of damping these vibrations and of designing a centrifuge and bearings which can tolerate the stresses of passing through these frequencies as the centrifuge is accelerated.

These problems were solved in the late 1950s by the use of a simple oil-lubricated pivot and cup bearing at the bottom and a magnetic bearing at the top [19b]. In the latter there is no physical contact between the rotor shaft and housing, and consequently no friction. The magnetic bearing which holds the shaft in suspension is also designed for damping vibrations, as is the lower cup and pivot bearing (see figure 6.5). Friction is further reduced by enclosing the centrifuge rotor in a casing which is maintained at very high vacuums, probably between one-millionth and one ten-millionth of an atmosphere [22]. The casing must not only be leak-proof but must also be strong enough to contain the debris of a failed centrifuge.

3. The very high rotational speed of the centrifuge leads to severe mechanical stresses in the outer wall. For example, the tensile stresses in the wall of an aluminium centrifuge with a radius of 10 cm and a rotational

Figure 6.5. A modern gas centrifuge

The thin-walled rotor is driven by a small electromagnetic motor attached to the bottom of the casing. The top end of the rotor is held in a vertical position by a magnetic bearing and does not touch stationary components. Gas is fed into and withdrawn from the rotor through the stationary centre post, which holds three concentric tubes for the feed, the product and the waste. The stationary bottom scoop protrudes into the spinning gas and provides a mechanical means for driving the vertical flow of gas. The top scoop, which serves to remove the enriched product, is protected from direct interaction with the rotating gas by the baffle, which has holes allowing the enriched gas to be bled into the area near the scoop. The baffle is needed to keep the top scoop from imposing a vertical flow that would counteract the crucial one generated by the bottom scoop.

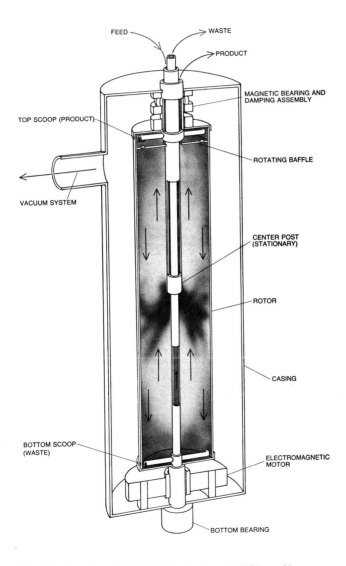

Source: *Scientific American*, Vol. 239, No. 2, August 1978, p. 29.

Table 6.1 Typical maximum peripheral speeds of thin-walled cylinders [20b]

Material	Tensile strength T (kg/cm^2)	Density ρ (g/cm^3)	$T\,\rho$ ($\times\ 10^3$)	Approximate maximum peripheral speed (m/s)
Aluminium alloy	5 200	2.8	1.9	425
Titanium	9 200	4.6	2.0	440
High-strength steel	17 000	8.0	2.1	455
Maraging steel	22 500	8.0	2.8	525
Glass fibre/resin	7 000	1.9	3.7	600
Carbon fibre/resin	8 500	1.7	5.0	700

frequency of 800 rev/s will be 7 000 kg/cm^2, more than 1.3 times the tensile strength of aluminium (see table 6.1). In other words an aluminium centrifuge would have exploded long before these rotational speeds were achieved. This problem has been solved by the creation of new high-strength, lightweight materials, such as carbon or glass fibres, some of the properties of which are compared with older materials in table 6.1. It should be emphasized that the peripheral speeds in the last column of the table are the 'bursting' speeds of the materials. Actual centrifuges made of these materials would be run at lower speeds to allow an ample safety margin and to prevent excessive distortion ('creep') of the rotor material over time.

4. The separative power of a centrifuge is very sensitive to the detailed nature of the countercurrent flow pattern. Until the hydrodynamic equations governing this pattern were understood it was only possible to improve centrifuge performance by empirical, trial-and-error methods. In recent years substantial progress has been made in developing computer codes for solving these very complicated equations [2t]. This has made it possible to optimize centrifuge designs in much more systematic and predictable ways.

With the solutions to these technical problems it has been possible to produce centrifuges with very high separation factors, at least 1.5, and with separative powers of anywhere from 5 to 100 kg SWU/yr [23a]. The large variation in this latter rating reflects substantial differences in the sizes and rotational speeds of the centrifuges. European and Japanese machines tend to be relatively small, while the United States has developed substantially larger machines with separative powers about 10 times those of the European and Japanese designs [2u].

The major factor limiting the separative power of a centrifuge is the very low throughput. This is limited by the slow diffusion rate in the gas and by the requirement that the pressure of the UF$_6$ at the centrifuge wall be below its sublimation vapour pressure at the operating temperature, usually normal room temperature, 20°C. If this latter condition is not satisfied, then solid UF_6 will deposit on the walls of the centrifuge, an obviously undesirable situation.

The vapour pressure of UF_6 at room temperature is only about 0.1 atmosphere [2v], and if this is taken to be the upper limit of the pressure at the wall, the pressure on the axis must be about 2.5×10^{-9} atmosphere, a very good vacuum. Using these limiting pressures and the exponential form for the density similar to equation 6.2, it can be shown that the total amount of uranium in a centrifuge 1.5 m long and 20 cm in diameter spinning at 800 rev/s is only about 0.25 g.

The separative power of a centrifuge is optimized by determining the proper ratio of product withdrawal to countercurrent flow, that is, the reflux ratio. With this criterion and certain assumptions about efficiency, it is possible to derive all the properties of a model centrifuge [24a]. Using these results the hypothetical centrifuge described in this section can be shown to have a separation factor of 1.51 and an optimum separative power of 15.2 kg SWU/yr. This puts it somewhat higher than the values typically quoted for European and Japanese centrifuges, but considerably lower than US values. The properties of this model centrifuge are summarized in table 6.2. It must be emphasized that these values are both hypothetical and approximate. They are hypothetical because the detailed operating characteristics of actual centrifuges are classified, and they are approximate because the formulae needed to derive them are accurate only for values of $g \ll 1$ [24b]. However, they should be close enough to realistic values to give a reasonably reliable description of the capabilities of a modern gas centrifuge.

Table 6.2. Properties of a hypothetical centrifuge

Radius	10 cm
Length	150 cm
Rotational frequency	800 rev/s
Peripheral speed	500 m/s
Separation factor (q)	1.51
Separative power	15.2 kg SWU/yr
Inventory	0.26 g U
Throughput	600 kg U/yr = 0.019 g U/s
Hold-up time	13.7 s

A centrifuge cascade

The values of table 6.2 can now be applied to the design of one of the cascades of table 5.1. For example, a commercial plant producing 3 per cent product with 0.2 per cent tails would require roughly 11 stages. Note that with $g = 0.51$ the approximation $g \ll 1$ is not really appropriate, and a more complicated calculation would have to be performed to get a better value. If the more precise $\ln q$ is used instead of g in table 5.1 to compute the number of stages the number is increased from 11 to 13.

Notice also that it is appropriate to use ideal cascade rather than square cascade values in table 5.1 because the use of many identical centrifuges allows the stages to be adjusted in width to approximate an ideal cascade (see figures 6.6(a) and (b)).

For a plant rated at 1 000 t SWU/yr about 66 000 of the centrifuges of table 6.2 would be required. The flow in the widest stage would be 5.9×10^6 kg U/yr, requiring that just under 10 000 centrifuges be placed in this stage. In practice the plant would probably be divided into modules, as is the Urenco facility at Almelo [25]. For example, a 1000 t SWU facility could be built up of ten 100 t modules, with each module being brought into production as it is completed. This kind of flexibility is one of the factors that makes the centrifuge so attractive compared to gaseous diffusion.

Another important advantage of centrifuges is their relatively low energy consumption. Quoted values range from 100 to 300 kWh/kg SWU, roughly a factor of 10 better than gaseous diffusion [23b]. If it is assumed that one of the centrifuges in table 6.2 uses 200 kWh/kg SWU, then the operating power of the centrifuge is only 350 W, and 66 000 such machines would require only a relatively small 23 MW power plant.

Finally, the equilibrium time of a commercial cascade is essentially negligible, only of the order of 2.5 minutes. It takes much longer than this to bring the centrifuges up to operating speed. The total plant inventory is also very small. The 0.26 g per centrifuge becomes a total of only 17 kg of uranium in the entire facility. This not only substantially reduces capital costs but also allows for much greater accuracy in accounting for material input and output. This has important implications for the application of safeguards to centrifuge plants (see chapter 3). The very short equilibrium time and small inventory suggest that a small centrifuge cascade might be easily used in a batch recycle mode to produce highly enriched uranium. However, this would require stopping and cleaning out all the centrifuges before restocking them with enriched feed material. This stopping and restarting of the centrifuges is both time-consuming and potentially damaging to the centrifuges, which must be accelerated through critical frequencies to bring them to operating speed. It would seem that commercial centrifuges are designed to suffer this experience only once and then to run at their nominal speed for many years without stopping [25]. Only systematic testing would determine how resistant they are to

Figure 6.6. Two centrifuge cascades
(a) Cascade hall at Almelo, the Netherlands

(b) Section of a Japanese cascade

repeated accelerations through the critical frequencies. Presumably such tests have been carried out, but the results are not available.

This discussion of the centrifuge can be summarized by stating that the technique seems to have reached maturity and is ready to compete on very favourable terms with gaseous diffusion. It is also clear that the centrifuge presents a qualitatively different and substantially more serious problem with respect to nuclear weapon proliferation than does gasous diffusion. This aspect of the centrifuge is examined further in chapter 2.

IV. Aerodynamic separation methods

Basic principles

There are a large number of techniques for separating isotopes which can be classified as 'aerodynamic'. A good definition which includes most of them is: " . . . aerodynamic separators are characterized as those involving preferential diffusion of disparate masses either driven principally by a pressure gradient generated by streamline curvature (Type I) or through molecular processes that involve large perturbations from an equilibrium distribution (Type II)" [26a].

The concept of diffusion across a streamline is quite analogous to the diffusion against a gravitational or centrifugal force which was discussed in the previous section. A streamline in a flowing fluid is a line across which no net material transport takes place. So all separation processes in the fluid occur in directions perpendicular to streamlines. For example, in a centrifuge the gas is moving in circular paths, so the streamlines are concentric circles. The isotope separation takes place in the radial direction, perpendicular to these lines. If a streamline is curved, this implies that the gas is being accelerated, and that a pressure gradient (i.e., force) must exist perpendicular to the streamline. This situation is common to all Type I aerodynamic processes, and both of the aerodynamic processes which have achieved or are approaching commercial viability (the German nozzle process and the South African advanced vortex tube process) are of Type I.

The large perturbations which characterize Type II processes can either be strong density or pressure gradients or the interaction between two rapidly moving jets of gas. These methods have been studied for some time under laboratory conditions, but so far none has been developed to even a prototype level for the separation of uranium isotopes. Economic studies based on early experimental data have generally concluded that the Type II aerodynamic methods are not likely to compete successfully with currently workable methods [27a, 28a]. Although experience warns that such assessments should be viewed with caution, there is at present no

136

compelling reason to consider these methods in detail in this book. However, a few will be described briefly in section VIII of this chapter.

The jet nozzle process

The aerodynamic nozzle process invented and developed by E. W. Becker and his associates has been the most successful to date of all the aerodynamic processes. In the Becker process a jet of gas consisting of roughly 96 per cent hydrogen and 4 per cent UF_6 is allowed to expand through a narrow slit [2w]. The gas moves at high speeds (comparable to those at the periphery of a modern centrifuge) parallel to a semicircular wall of very small radius of curvature (see figure 6.7). If the speed of the gas is 400 m/s, and the radius of curvature is 0.1 mm, then the centrifugal acceleration achieved is 1.6×10^9 m/s^2 or 160 million times gravity. The accelerations exceed even the high values achieved in centrifuges by a factor of a thousand or more, and they are achieved in an apparatus with no moving parts. The centrifugal forces on the molecules cause the streamlines of the heavier components of the gas to move closer to the curved wall than those of the lighter components as the gas flows around the semicircle. At the other side, where the gas has changed direction by 180°, a sharp 'skimmer' separates the flow into an inner light fraction and an outer heavy fraction.

Figure 6.7. A separation nozzle

A cross section of the separation nozzle system currently in use. The knife edge skimmer at the right is placed so that 25 per cent of the UF_6 in the feed goes into the light fraction and 75 per cent into the heavy fraction.

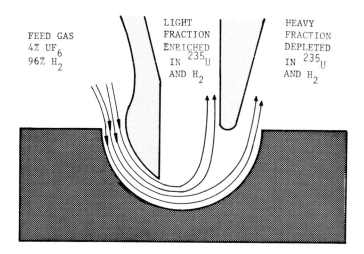

137

Figure 6.8. A separation nozzle element
(a)   Stack of photoetched metal foils forming a separation nozzle structure. The diameter of the curved deflection wall is 0.2 mm

(b)   Assembly of separation nozzle elements by stacking metal foils into chips which are then set into a tube. A jet nozzle stage comprises a large number of such tubes (see figure 6.10)

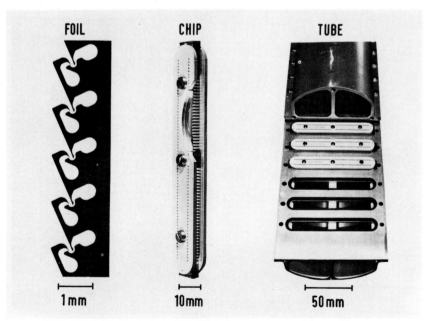

The position of the skimmer is arranged so that one-quarter of the total UF_6 content is extracted in the light fraction. This is the 'cut'. The other three-quarters in the heavy fraction is depleted in $^{235}UF_6$ and forms the tails from the separating element.

The hydrogen gas plays two crucial roles in amplifying the separative effect of the nozzle. First, its low molecular weight greatly decreases the average molecular weight of the process gas. This increases the sonic velocity of the gas and allows the flow through the nozzle to be subsonic. This eliminates shock effects which would otherwise absorb energy and create turbulence and some remixing of the separated isotopes. The UF_6 molecules are dragged along by the fast-moving hydrogen and achieve much greater peripheral speeds than could be achieved in pure UF_6 gas.

The second beneficial effect of the hydrogen is also the result of drag forces it exerts on UF_6 molecules. As the latter move around the curve they tend to 'fall' to the outside, and if it were not for the hydrogen they would fall at the same rate. However, the drag forces exerted by the hydrogen act unequally on the light and heavy isotopes and cause the $^{235}UF_6$ molecules to fall more slowly than the $^{238}UF_6$. This is a non-equilibrium effect, which substantially enhances the separation as long as the skimmer takes its cut before equilibrium is achieved [29].

The extremely small size of the separative nozzle allows the process to operate at relatively high pressures and velocities and to also have the gas flow 'laminar', or non-turbulent. In fluid mechanics an index called the Reynolds number is used to indicate the transition from laminar to turbulent flow, and the transition is at a value of around 2 000. The Reynolds number is proportional to the linear dimensions of the nozzle, and in order to keep the value at about 100, well under the onset of turbulence, the slit can be only about 0.05 mm wide or about one-half of the curvature radius (see figures 6.8(a) and (b)). This implies that the rate of material flow must be very small in a single nozzle, and that many thousands of nozzles are necessary to make up an enrichment stage of reasonable size.

The tiny nozzle elements are produced on thin metal foils by a photoetching process [2x]. The foils are then stacked to make elements or chips with slit lengths of a few millimetres and these are enclosed in two cover plates. One cover plate has rows of holes which open into the feed chambers of all the elements and the other cover plate has holes which connect to the tails chambers. The light, enriched fraction emerges from the sides of the strip. Then 80 chips are mounted in a tube which is two metres long. The feed is admitted to the chamber on one side of the tube and the tails extracted from the other. The enriched product emerges from the space between the two tube halves. It is claimed that such a tube can produce a separation gain of 0.0148 with a cut of 0.25, and a separative work capacity of 50 kg SWU/yr [30].

Before these data can be used to construct a cascade the cascade theory developed in chapter 5 must be modified to take account of the

Figure 6.9. An asymmetric cascade

A cascade assembled from elements with a cut of one-quarter. Product from stage n is sent forward to stage $n + 3$ while tails are sent back one stage. Note the three product streams proceeding upward in parallel.

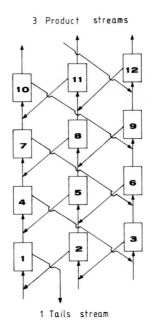

3 Product streams

1 Tails stream

asymmetrical cut of the nozzle process. The essential criterion that must be satisfied is that streams of differing concentrations are not mixed. Applying this criterion leads to a set of equations for an ideal asymmetric cascade whose solutions are similar in form to those of the ideal symmetric cascade but which require a different arrangement of process streams [2y]. In particular, if the cut is $1/n$ then the product stream from a stage must be sent forwards $n-1$ stages and the tails stream backwards one stage. When $n = 4$, which corresponds to the nozzle process, this results in a cascade like that of figure 6.9. The figure shows that in an asymmetric cascade there are several process streams moving up the cascade in parallel. As a general rule for a cut of $\theta = 1/n$ there will be $n-1$ such streams.

If the separation factor of the stages is close to 1, then a full analysis of the asymmetric cascade shows that all of the formulae of chapter 5 and table 5.1 can be used directly, with the following two adjustments [2z].

1. Product flow P must be taken as the sum of the product flows from all parallel streams.

2. The separation gain g is to be replaced by $2\theta g$.

As an example consider a jet nozzle facility for production of 3 per cent product and 0.2 per cent tails. With $g = 0.0148$ and $\theta = 0.25$ the number of stages in an ideal cascade comes out to be 743. For a square cascade the result is 649. This can be compared with the data given for the

Figure 6.10. A prototype jet nozzle stage

Feed enters from the right centre and passes through the separation elements arranged around the top chamber. The light, enriched fraction is drawn down through a heat exchanger and compressor and sent on to another stage. This test prototype has the tails being recycled directly and mixed with the feed.

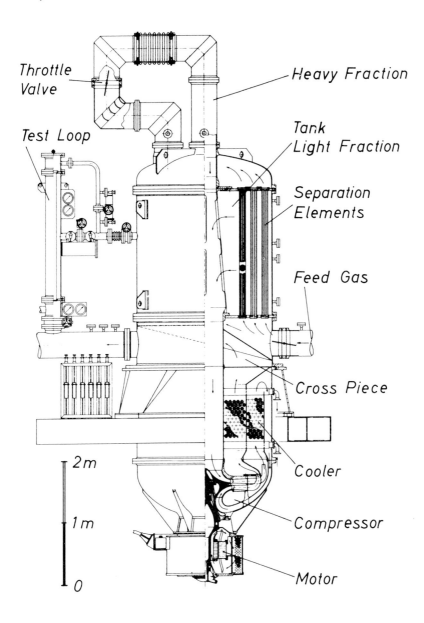

Source: Becker, E.W. *et al.*, *Present State and Development Potential – Separation Nozzle Process*, KFK Report 2067 (Institut für Kernverfahrenstechnik, Gesellschaft für Kernforschung, Karlsruhe, FRG, 1974).

conceptual design of a large enrichment plant based on the nozzle process and designed to produce 3 per cent product with 0.34 per cent tails [15b]. This is a squared-off cascade having two different sizes of cascade element and containing a total of 570 stages. Considering the higher tails assay, this compares quite well with the estimate in table 5.1.

The higher than usual tails assay quoted for this demonstration plant is almost certainly connected to the high energy costs of SWUs in a nozzle plant. The pumping and pressurizing of large quantities of gas demands a great deal of energy, even more than for gaseous diffusion. This results mainly from the fact that only 4 per cent of the process gas is UF_6. The energy costs for the nozzle process are generally quoted as between 3 000 and 3 500 kWh/SWU [23b]. As was shown in chapter 5 such high energy costs could dictate higher tails assays, since the cost of the extra ^{235}U thrown away could be less than that of the energy which would be needed to retrieve it. Whether or not this will be true in full-sized commercial jet nozzle plants remains to be seen.

Figure 6.10 shows a prototype stage element. It has many features in common with a gaseous diffusion stage including a compressor, heat exchanger and 'diffuser' section. It is in the latter that the actual separation takes place. Given these similarities it seems likely that the hold-up time for a jet nozzle stage is comparable to or possibly somewhat less than that of a gaseous diffusion stage. In the absence of any published information on hold-up time it will be estimated that the value is between 1 and 5 seconds.[2] Since the effective separation gain ($2\theta g = 0.0074$) is larger than that for gaseous diffusion, and the hold-up time possibly somewhat shorter, the inventory requirements and equilibrium time for a nozzle cascade should be lower than for gaseous diffusion.

One special problem which arises in the aerodynamic processes is the handling of hydrogen. As was mentioned in chapter 5, UF_6 reacts strongly with hydrogen at elevated temperatures, so care must be taken to keep this reaction from occurring. In addition, the separation nozzles tend to separate the hydrogen from the UF_6. It turns out that this problem is not serious when the elements are connected in a cascade, presumably because the product stream which has been mildly enriched in $^{235}UF_6$ but strongly enriched in H_2, is mixed at the entrance to the next stage with a tails stream which has been depleted in H_2. But there is still a problem at the product outlet of the cascade where the hydrogen has to be purified and recycled to the bottom of the cascade [2aa].

An interesting recent development which will be applied in the new Brazilian facility is the use of a more elaborate nozzle system to divide the gas into three fractions instead of two (see figure 6.11) [32]. The intermediate fraction is then recycled within the stage in a form of internal reflux. Just as in other uses of reflux, this has the effect of increasing the

---

[2] A hold-up time of two seconds has been given by one source [31].

Figure 6.11. Advanced separation nozzle

This nozzle system employs a double deflection of the jet to produce a light, intermediate and heavy fraction. The intermediate fraction is recycled within the stage as an internal reflux.

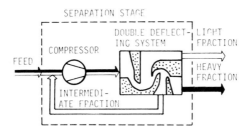

separation gain while reducing the throughput, but optimization of the reflux ratio can produce significantly greater separative power in a given stage element. It also changes the optimum cut from one-quarter to one-third, reducing the number of stages needed.

The Helikon process

The most succinct description of the South African enrichment process has been the following: "The UCOR process may best be described as a combination of a stationary-wall centrifuge with a highly asymmetric cut and a cascade system (the helikon system), which is eminently suited to asymmetric separation" [13c]. The 'stationary-wall centrifuge' is thought to be related to a device called the vortex tube [26b]. Just how close the relation is is difficult to judge, especially since a group of the project's workers has said that while the separating element was originally based on the vortex tube its present form is "far removed" from it [13d].

With this caveat in mind it may still be useful briefly to examine the working of a vortex tube. This is a tube into which a mixture of $1-2$ per cent UF_6 and $98-99$ per cent H_2 gases [33] is injected at high velocity tangential to an inner wall, but with an axial component of velocity as well (see figure 6.12). As the gas spirals around the tube the lighter component tends to concentrate near the axis just as it does in the rotating centrifuge or jet nozzle. When the gas reaches the other end of the tube the outer and inner portions are drawn off separately.

Very little work on isotope separation with the vortex tube has been done outside South Africa, and one early assessment of the process was quite negative [27b]. However, this study apparently underestimated the improved performance which can be obtained by diluting the UF_6 with a light carrier gas such as H_2. The same enhancing effects which occur in the jet nozzle process also seem to operate in the vortex tube to give greatly improved separation factors over what can be obtained with pure UF_6. The

143

Figure 6.12. A vortex tube

A mixture of UF_6 and H_2 is injected tangent to the tube's inner wall and spirals down the tube. At the end the heavy and light fractions are separated by a skimmer. In this tube the cut is only one-twentieth.

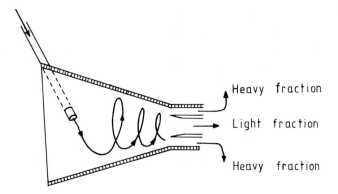

separation factor most often quoted for the South African process is $q = 1.03$ [34a].

However, this large separation factor can be obtained only at the price of taking a very small cut, $\theta = 0.05$ [34b]. This means that 19 separate streams of product, each with a slightly different enrichment, must be moved forward without mixing. Such a process is economically prohibitive if separate transmission pipes and compressors must be provided for every stream.

A very imaginative solution to this problem has been developed by South African researchers, who discovered that many separate streams can be sent simultaneously through the same axial-flow compressor without significant mixing among them. An axial-flow compressor consists of a rapidly rotating shaft driving a series of blades similar to turbine blades (see figure 6.4). These are mounted in a casing to which are attached arrays of stationary blades which alternate with the rotating ones. As gas is drawn into the front of the compressor it is alternatively speeded up by the rotating blades and deflected by the stationary ones in such a way that it emerges from the back of the compressor at a higher pressure. The most familiar application of such compressors is in jet engines, and it was on a jet engine that the South Africans did their original experiments on flow mixing [34b].

The use of this technique allows the combining of all 19 elements of a given cascade level into a single module utilizing only two compressors. This compact design, coupled with the relatively high separation gain associated with the small cut, allows the design of a commercial-type cascade with only 100 modules [34a]. It should be emphasized that these 100 modules actually constitute 1 900 enrichment stages. This is consistent with a separation gain of $2\theta g = 0.0026$, about two-thirds of the value

144

obtainable in gaseous diffusion. One important advantage claimed for the modules is that they are flexible in the way the various streams are organized. This allows for a squared-off cascade to be built up entirely of identical modules with only the connections being altered to allow for changes in flow rate and enrichment factor [34c].

A single module would contain two axial-flow compressors with associated heat exchangers, the separation elements and the piping necessary to move the 19 feed streams into and out of the module while the single tails stream is deflected within the module through each stage in succession (see figure 6.13). A module with a capacity of 10 t SWU/yr would be about 3.6 m in diameter, and if figure 6.13 is roughly to scale, about 10 m long [13d]. Larger modules with diameters of 6.5 m have been proposed and have been rated at 50−65 t SWU/yr [28b]. Recent improvements are said to have raised this to between 80 and 90 t SWU/yr for a module of the same size [13e]. These modules would be comparable in size to the largest gaseous diffusion stages, but only about one-tenth as many would be needed to make a plant of the same capacity.

The energy consumption of the South African process is comparable to that of the jet nozzle, somewhere in the range 3 000−3 500 kWh/SWU [13f]. The equilibrium time is stated to be low, only 16 hours for a plant producing 3 per cent product [28b]. Using table 5.1 and the value $2\theta g = 0.003$ gives a stage hold-up time of between 0.05 and 0.2 seconds. This is the hold-up time per *stage*, and since there are 19 stages in a module, the modular hold-up time is probably an order of magnitude

Figure 6.13. A prototype Helikon module

This module is 3.6 m in diameter and is rated at 6−10 t SWU/yr. It contains two axial flow compressors with associated heat exchangers. The central region contains the separating elements and the piping to route the tails stream through the stages in a helical pattern. The drawing does not illustrate these details clearly.

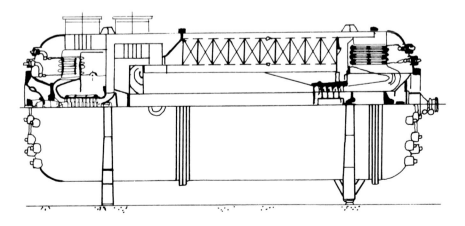

Source: Roux *et al.*, in *Nuclear Power and Its Fuel Cycle* (IAEA, Vienna, 1977), p. 178.

larger. The small hold-up time also leads to a small inventory, making the Helikon process a relatively convenient one for batch recycling to produce higher enrichments. This may be the means by which South Africa was able to produce uranium enriched to 45 per cent for its research reactor at Pelindaba, an achievement which was announced on 29 April 1981 [35].

V. Chemical-exchange methods

Basic principles

The chemical-exchange process is an excellent object lesson in the difficulty of making reliable predictions about the development of enrichment processes. Many attempts were made in the early years after World War II to find chemical enrichment techniques, but in all cases the methods were found to be far too slow and expensive to compete with gaseous diffusion [36]. Although chemical-exchange techniques had proved very successful in separating lighter isotopes, it was known that the separating effect decreases with isotopic mass. This implied that chemical separation of uranium was likely always to be difficult, and that finding a successful process could require substantial further research in uranium chemistry. This was the general situation described in a review of these methods in 1972 [27c].

However, in 1968 French researchers had discovered a new technique [37], and after nine years of development the process was announced at the 1977 IAEA Conference at Salzburg [13g]. More recently the Japanese have announced a successful process developed by the Asahi Chemical Company, which is currently planning construction of a pilot plant [38a]. So, in a very short period of time chemical enrichment of uranium has become a prominent candidate for future development, one with some important implications for the proliferation issue.

The chemical-exchange method of isotope separation depends on the very small tendency of different isotopes of an element to concentrate in different molecules when there is an opportunity for exchange between the molecules. As an illustration consider two uranium compounds AU and BU which are mixed together and allowed to come to equilibrium. In general each compound will have two isotopic forms: $A^{235}U$ and $A^{238}U$; and $B^{235}U$ and $B^{238}U$. Therefore, in any mixture of the two uranium compounds there will be four different species present. If these species are chosen properly, they can exchange uranium atoms. Chemists depict such an exchange equilibrium in the form of an equation

$$A^{238}U + B^{235}U \rightleftharpoons A^{235}U + B^{238}U$$

where the double arrows imply that the reaction can proceed in both directions.

When the system is in equilibrium each species is present in a certain concentration, usually measured in moles per litre and denoted by the symbol [AU]. The equilibrium is characterized by an equilibrium constant K which gives the relation among the four concentrations

$$K = [A^{235}U] [B^{238}U] / [A^{238}U] [B^{235}U] \qquad (6.6)$$

The chemical equation is customarily written to make K greater than 1, so this implies that in equilibrium the product of concentrations with ^{235}U in A and ^{238}U in B is greater than the product with ^{238}U and ^{235}U interchanged. Therefore if the two compounds AU and BU initially contain identical isotopic ratios, after the equilibrium is established compound A will be slightly enriched and B slightly depleted in ^{235}U.

Unfortunately it is not easy to explain why the value of K is different from 1. Indeed, the equipartition theorem of statistical mechanics, which was used to explain the separation effect in gaseous diffusion, predicts that K should always equal 1, and that no isotope separation should occur. The fact that it does occur is explained by the theory of quantum mechanics and the connection between the energy of a molecule and its vibrational frequencies. No such connection exists in the older classical physics.

The quantum theory states that the unit of vibrational energy of a molecule is proportional to its frequency, and that the molecule can only absorb energy in amounts which are a multiple of this basic unit. This means that the classical equipartition theorem cannot be valid for such a system, since the theorem requires that any amount of energy, no matter how small, which is added to the system must be shared equally among all the molecules. This situation exists to a good approximation if the temperature is high, but as the temperature is reduced the equipartition principle breaks down. In practice this results in the lighter isotope tending to concentrate in the more loosely bound molecule, while the heavy isotope is more likely to be found in the tightly bound molecule [11b].

This explanation immediately suggests two requirements on the compounds for a good chemical-exchange separation:

1. The two compounds should be very different, with the uranium tightly bound in one and loosely bound in the other. Free uranium ions fit the second requirement best of all.

2. The equilibrium should be established at the lowest possible temperature to maximize the separation effect.

Unfortunately, both of these requirements are incompatible with the requirement of easy exchange of the uranium. Two very different compounds do not exchange components effectively, and reducing the temperature reduces the rate at which equilibrium can be established.

147

This problem can be dealt with by using suitable catalysts. These are compounds or materials which do not participate directly in the chemical reaction but whose presence in some way facilitates or speeds up the reaction. The ultimate success of the French and Japanese chemical enrichment processes seems to have been largely the result of the discovery of suitable catalysts. These will be discussed in more detail below.

The equilibrium constant K of equation 6.6 is identical to the single-stage separation factor q. This can be demonstrated as follows. Consider a stage to be a single operation of mixing the two compounds AU and BU, with both having equal proportions of light and heavy uranium isotopes. After equilibrium is achieved the compound AU will be slightly enriched in the light isotope and BU will be slightly depleted. Compound AU can therefore be called the product, and BU the tails coming out of this stage. .

The relative isotopic abundances R_P and R_W in the two compounds are simply the ratios of the concentrations of the two isotopic species in each compound:

$$R_P = [A^{235}U] / [A^{238}U]$$
$$R_W = [B^{235}U] / [B^{238}U]$$

(6.7)

and the separation factor is given as in chapter 5 (p. 95) by

$$q = R_P / R_W$$

(6.8)

Inserting equations 6.7 into 6.8 and rearranging the factors reproduces equation 6.6 and shows that

$$q = K$$

(6.9)

This relationship holds as long as each compound contains only one atom of uranium. In more complicated compounds the formula must be modified, but such refinements are not needed for a basic understanding [18b].

For isotopes of light elements the value of q can be quite large, and in fact isotopes of hydrogen, boron and nitrogen are most effectively separated by chemical means [18c]. For uranium, however, q is limited to values at or below 1.003, and a great many stages are required to achieve useful enrichments. For example, if $q = 1.002$, over 2 700 stages are required in an ideal cascade to produce 3 per cent enriched product with 0.2 per cent tails. But chemical separation 'stages' are quite a bit simpler than the other types discussed so far, and this number is not as prohibitive as it may seem.

A chemical separation stage consists simply of the thorough mixing of two substances for a sufficient time to allow chemical equilibrium to be established. This is most efficiently done in a device called a 'counter-current column' in which one of the compounds is carried upward and the other downward. Any level in the column is associated with a given isotopic composition of both compounds, so that only compounds of the same composition are brought into contact. The very slight isotopic transfer causes one compound to be slightly enriched and the other depleted as each moves on to the next contact.

For an efficient exchange the two compounds AU and BU must be both easily brought together and easily separated. This is accomplished most efficiently if the two compounds are in different phases. One might be a liquid and the other a gas; or both could be liquids which are immiscible; or one could be a liquid or gas and the other a solid. The liquid–gas system operates very much like a standard fractional distillation column, the liquid–liquid system like a solvent extraction process, and the liquid–solid system works like an ion-exchange column or 'chromatography'. All of these techniques are widely used and very well understood by the petroleum and chemical industries, and this great wealth of experience adds to the attractiveness of chemical separation. Another attractive feature is the potentially very low energy consumption of the process. Since chemical exchange is an equilibrium process there is no need for powerful compressors or pumps for the preparation of phases for contacting.

These advantages are counterbalanced by the problem of processing the two compounds at the ends of the column, the problem of 'reflux'. To illustrate this problem, suppose that compound A is moving upward in the column and B downward (see figure 6.14). This means that the highest enrichment in ^{235}U is at the top, and the enriched uranium emerges from the top in the compound AU. Some of this can be removed as product (or sent on to another column), but most of it must be refluxed to maintain the large countercurrent flow. However, before it can be sent back down the column the AU must be converted to BU. (The opposite must be done at the lower end, the tails reflux.) This is in general a non-reversible chemical process, which can require large amounts of energy and which must be done very carefully to avoid losses of enriched uranium. Because the reflux ratio is generally very large (i.e., the amount recycled is many times the amount extracted), even a small percentage loss of material in the reflux reaction can significantly reduce the efficiency of a plant (see chapter 5, p. 115).

With this general introduction to the theory of chemical enrichment it is now possible to consider the two most promising processes in more detail. The first is the Japanese process which depends on an exchange of uranium between a liquid solution and a finely divided ion-exchange resin. The second is the French solvent extraction process which uses an exchange between two immiscible liquid phases, one aqueous and the other organic. A third process, studied extensively in the USA, but apparently considerably less developed than the first two, is also a liquid–liquid process based on exchange between solutions of UF_6 and $NOUF_6$.

Figure 6.14. A countercurrent column

The compounds AU and BU are repeatedly mixed and separated as they move up and down the column respectively. The product reflux chemically converts AU to BU while the tails reflux converts BU to AU.

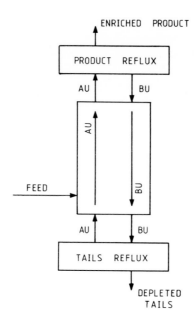

The Japanese process

In this process a column or series of columns is packed with an ion-exchange resin analogous to those used in many industrial purification processes or in domestic water softeners. This resin has the property of attracting and holding to its surface (adsorbing) some chemical species in preference to others, so it serves as the medium or 'phase' for holding one of the two uranium compounds to be contacted. The resin is in the form of a fine powder in which the individual particles are spheres of only 20−200 μm in diameter [39]. This provides a large area for contacting the two phases, and it is quite analogous to the function performed by a gaseous diffusion barrier.

The resin is first prepared for receiving the uranium compound by charging it with a strong oxidizing agent, called, for convenience, Ox^{II} [40a]. When the resin is saturated with the oxidizing agent, a new solution is introduced containing a uranium compound in which the uranium is in the IV oxidation state (see chapter 5, p. 119). When this comes into contact with the oxidizing agent on the resin, it is oxidized to U(VI) and replaces the now reduced oxidizing agent on the resin. This reaction is quite rapid and leads to a well-defined boundary between resin saturated in Ox^{II} and in U(VI) (see figure 6.15). This boundary moves slowly down the column as more U(IV) solution is added at the top. When an appropriate length of resin has been saturated with U(VI), a new chemical solution is introduced at the top of the column. This is a reducing agent, referred to only as Re[I] [40b], which acts on the U(VI) adsorbed on the resin to reduce it to U(IV), and also to replace it on the resin. The released U(IV) compound goes into solution and migrates down the column, coming into contact with the

Figure 6.15. An ion-exchange module
The four columns on the left are filled with ion-exchange resin particles. In the regions marked "RED" and "OX", the resin is in its reduced and oxidized states respectively, and no uranium compounds are adsorbed on it. In the regions marked "U", a U(VI) compound is adsorbed on the resin and is in continuous contact with a U(IV) compound in solution. In the fourth column the reduced resin is being oxidized in preparation for the arrival of the leading edge of the "U" column. Rotating valves at the tops and bottoms of the columns control the flow of oxidizing and reducing agents and uranium compounds. They are also programmed to admit feed and remove product and tails at appropriate times. These arrangements allow continuous operation of the module.
The two columns on the right are used to recharge and recycle the reducing and oxidizing agents. The input of "O_2" symbolizes reoxidation of the oxidizing agent while "H_2" symbolizes reduction of the reducing agent. These symbols do not necessarily imply that oxygen and hydrogen are the actual substances used in the recharging process.

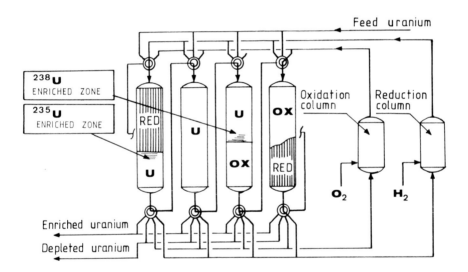

Source: M. Seko et al., Nuclear Technology, Vol. 50, No. 2, September 1980, p. 182.

U(VI) still adsorbed further down. In this region the isotopic exchange reaction

$$^{235}U(IV) + {}^{238}U(VI) \rightleftharpoons {}^{238}U(IV) + {}^{235}U(VI)$$

can take place, resulting in a slight concentration of ^{235}U in the U(VI) compound adsorbed on the resin. The equilibrium constant usually quoted for this reaction is 1.0013 [40c]. This is therefore the single-stage separation factor for the process (see equation 6.9).

The reducing agent at the top of the column also creates a sharp boundary, so that the portion of the column in which the isotope exchange takes place is a well-defined, slowly descending segment. The length of this segment is determined by the desired degree of enrichment and the flow rate of the solution through the resin. Enrichment in ^{235}U occurs at the top of the segment, since the ^{235}U tends to concentrate in the U(VI) compound which remains adsorbed on the resin.

The segment of column between the two boundaries moves slowly downward, but it is easier to analyse if viewed from a frame of reference moving at the same speed as the boundaries. In this frame of reference the resin and its adsorbed U(VI) are moving upward; the solvent and its dissolved U(IV) are moving downward; and reflux is occurring at both ends. This makes clear the countercurrent nature of the process and permits analysis of the system by the standard methods applied to any countercurrent square cascade [40d]. Therefore the formulae derived in chapter 5 (see table 5.1) can be used directly to estimate the important parameters of the system. The value of q can be taken to be 1.0013 and the cut assumed to be equal to a half. The total number of stages required for 3 per cent product, 0.2 per cent tails would therefore be 3 690 (an ion-exchange column must be treated as a square cascade) [38b].

The definition of a stage in this process is not as obvious as in those previously described. The flow of materials is continuous, so a stage is a more abstract concept similar to the notion of 'theoretical plate' in fractional distillation [11c]. The length of a stage in chemical exchange is determined by the rate of flow of materials and the equilibrium time of the reaction. Recent statements give the equilibrium time as "less than a second" [40e], and the thickness of a stage can be estimated from the statement that "several hundred theoretical separation stages per meter" are achievable [38b]. So a stage will be only a few millimetres long and the velocity of flow will be of the order of millimetres per second. A typical U(IV) molecule would therefore take roughly one hour to migrate from the top of the exchange band to the bottom. Note that this velocity has no simple connection to the rate at which the exchange region boundaries move.

The key to making this process commercially attractive has been the reduction of the equilibrium time by the use of catalysts, in particular resins which lowered previous reaction times by a factor of 1 000 [38c]. It also required many years of research to find suitable oxidizing and reducing agents as well as economical methods of recharging and recycling them. Considering the fact that quite negative assessments were being made of this method as recently as 1975 [15c], this example can serve as a useful reminder that even the most frustrating technological problems have a way of being solved if sufficient incentive is present.

As figure 6.15 shows, this process does not use a single column, but is carried out in a series of three columns while the resin is being recharged in the fourth. By programming the valves at the top and bottom of the column the moving liquid column can be recycled indefinitely through the four-column module. As the leading edge of the uranium band passes a valve, a quantity of tails can be removed and, similarly, product can be removed and feed added as the appropriate portions of the band pass through the valves. All of this has apparently been achieved by the design of appropriate valves, detectors for determining separation band boundaries, and computerized control systems [40f].

The amounts of material and sizes of equipment involved can also be estimated with the help of table 5.1. If the exchange band is 'several' metres long, figure 6.15 suggests that each column should be a 'few' metres in height. From table 5.1 it can be shown that the total inventory of uranium in the enrichment band is

$$I = 44 \ t_h/g^2 \tag{6.10}$$

Note that for this process the values for a square cascade are the most appropriate, since the column has a constant diameter. Taking $t_h = 1$ s and $g = 0.0013$ in equation 6.10 gives an inventory of 2.6×10^7 kg U for every kg/s of product, and since the latter represents 4.3 kg SWU/s the specific inventory can be shown to be 0.19 kg U/kg SWU/yr. These values are roughly twice the corresponding number for gaseous diffusion (see p. 133). This means that an ion-exchange plant will necessarily have a large inventory relative to its production rate, but it should also be pointed out that ion-exchange facilities need not be as large as gaseous diffusion plants to be economically viable [38d].

A rough idea of the size of a column can be obtained from the inventory values. Consider a four-column module with a capacity of 1 000 kg SWU/yr designed for 3 per cent product and 0.2 per cent tails. Such a module would contain 190 kg of uranium, and if this is present in solution with a concentration of the order of 0.1 mole/litre [40b], a total of about 8 000 litres of liquid must be present. If this is in the form of a column 4 m long then the diameter would be 1.6 m. Allowing for the volume taken up by the resin particles one might guess that a typical 1 t SWU/yr module would consist of four columns each about 3 m high and something under 2 m in diameter. A 1 000 t SWU/yr plant would have 1 000 such modules, or fewer larger ones.

Since the chemical-exchange process is reversible, very little energy is required for the ion-exchange columns (unless they must be kept at an elevated temperature, but this is not indicated in the most recent literature). The main energy expenditure comes at the regeneration of the oxidizing and reducing agents and in the pumping of recycled solutions. It is not possible to estimate the specific energy consumption without more data [40g], but it seems reasonable to assume that it is considerably less than gaseous diffusion, but probably somewhat larger than the centrifuge.

The French process

This process utilizes chemical exchange between two uranium compounds dissolved in two immiscible liquids. One of the phases is aqueous and the other an organic solution, and they are made to flow through each other in a countercurrent column. The contacting of the two phases is achieved by agitating them so that the organic liquid breaks up into small droplets suspended in the aqueous solution. The effect is quite similar to that achieved by shaking a bottle of oil and vinegar salad dressing. In an industrial separation column this agitation can be achieved in several possible ways, but the French process employs a "pulsed-column" [37]. This is a column in which the contents are agitated either by an external mechanical device or by a series of reciprocating discs installed in the column [41a].

The isotope exchange rate is limited by two important factors: the inherent rate of the chemical reaction and the diffusion rate of the two chemical species into and out of the volume of the oil droplets. The former can be speeded up by an appropriate choice of compounds and by catalysts and/or temperature increases. The key to the success of this process seems to be the use of an unusual kind of compound called a 'crown ether' [42a]. An example of a crown ether, dibenzo-18-crown-6, is shown in figure 6.16 [43]. The name 'crown' was given to this compound because of the shape of the molecule, which in three dimensions has the oxygen atoms arranged in a circle above the ring much like the points of a crown.

The crown ether is part of a larger class of compounds called chelating agents which are very useful in many branches of chemistry [41b]. These

Figure 6.16. A molecule of dibenzo-18-crown-6

This crown ether has 18 atoms in its basic ring structure, six of which are oxygens. The structures on each end are benzene rings, each containing six carbon atoms and associated hydrogens.

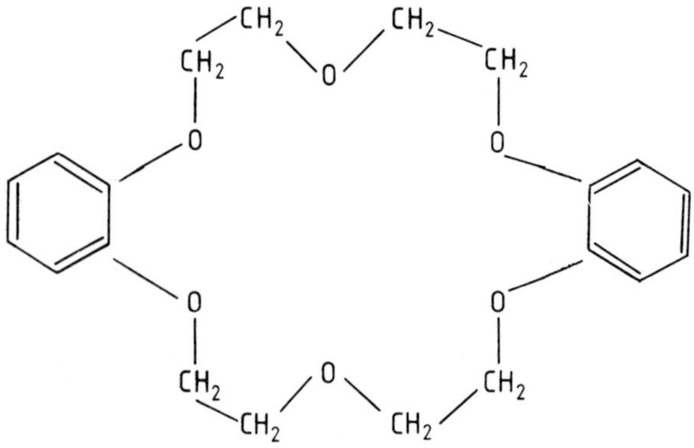

agents work by forming co-ordination compounds with metal ions. For example, the molecule of figure 6.16 will very readily react with potassium ions so that each of the oxygen atoms in the ring contributes a pair of electrons to a co-ordination bond with the potassium ion. The metal ion is then said to be 'sequestered', and in this form it can either be inhibited from undergoing reactions it might otherwise participate in, or encouraged to do things it would not otherwise do, for example, to dissolve in an organic solvent [41c].

Crown ethers can be made in a variety of sizes, and the key to making a crown ether effective for a particular ion is to match the size of the opening in the crown to the size of the ion in solution. Presumably this is one of the discoveries which underlies the successful French process. If a crown ether has been made which forms a co-ordination compound with uranium ions, then this compound can serve the dual function of extracting the uranium from one liquid phase to another and enhancing the separation effect. This latter property follows from the fact that the separation effect is directly proportional to the number of co-ordination bonds connected to the uranium [44].

The second factor which limits the exchange rate is the rate of diffusion of uranium compounds into and out of the oil droplets. This can be increased by making the droplets as small as possible, just as the ion-exchange reaction is enhanced by using very small resin particles. However, it does not seem to be feasible to agitate the two liquids violently enough to produce droplets smaller than a few millimetres in diameter [45a]. This factor seems to be the main one in making the stage hold-up times in the French process at least 20 times as long as those in the Japanese process, that is, of the order of $20-30$ seconds [45b].

The separation gain has been given as either "greater than 2×10^{-3}" [46a], or "about twice higher than the best ones [previously] known" [45c]. Given that values of 0.0013 to 0.0016 are well known for other processes [38b, 47], it is reasonable to speculate that a value between 0.0025 and 0.0030 has been achieved in the French process. This suggests that a square cascade designed to produce 3 per cent product and 0.2 per cent tails should have between 1 600 and 1 900 stages. If the total height of the reaction column is taken to be 30 m (see below) then a single stage has a length of between 1.6 and 1.9 cm. Note that both the stage height and column height are about an order of magnitude larger than those of the Japanese process. The lower separation factor of the latter seems to be more than compensated for by its much faster equilibrium time.

The proposed French enrichment facility would be constructed of units each consisting of two identical columns connected as shown in figure 6.17 [48a]. Note that this mode of connection has the effect of turning two square columns into a squared-off cascade. This increases the cascade efficiency (i.e., the approximation to an ideal cascade) to about 85 per cent, thereby reducing somewhat the very long equilibrium time [48b].

Figure 6.17 A chemex unit

Each unit consists of two columns connected as shown to form a squared-off cascade. Feed material is supplied to each column at the appropriate level. The columns of such a unit would be about 20 m tall and 1 m in diameter. An industrial module with a capacity of 200 t SWU/yr would comprise 35 such units or 70 columns.

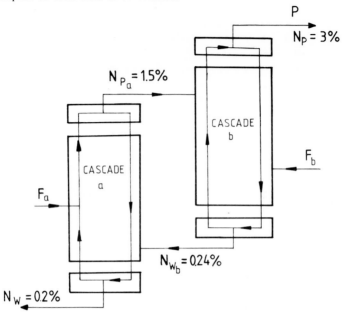

P = product; N_p = product assay; N_{p_a} = a cascade product assay; N_{w_b} = b cascade tails assay; W = tails; N_w = tails assay; F_a, F_b = feeds of natural uranium.

Each column will be more than 20 m high and more than 1m in diameter (a volume of 15.7 m^3) and will be capable of producing 3 000–5 000 kg SWU/yr [48c]. This implies a product flow rate of 1 400–2 350 kg/yr from a two-column unit. Using a hold-up time of 25 s (see below) and g = 0.0028 in table 5.1 the total uranium inventory in the two columns can be estimated as between 6 and 10 t. Using the total volume of 31.4 m^3 the average concentration of the uranium solutions is found to be roughly 1 mol/l, considerably more concentrated than those employed in the Japanese process. This makes for a very large uranium inventory, acknowledged to be an order of magnitude larger than that for a gaseous diffusion plant of similar capacity [48b]. This same ratio holds approximately for the equilibrium times as well, with a Chemex plant requiring 15 months to begin producing 3 per cent enriched product [46b]. This implies a hold-up time of 20–30 s.

It is not only the inventory of uranium which is going to be large in a Chemex plant. The volumes of all other necessary chemicals must also be correspondingly large, and some of them, such as the crown ethers, may be relatively expensive. There is also the problem of refluxing these enormous quantities of chemicals with very high efficiency. This problem is claimed to have been solved, and 'established technology' has been adopted wherever possible. The energy consumption is claimed to be less than 600 kWh/kg SWU [37].

The only secrets in the process seem to be the uranium compounds themselves, the catalysts, and the specific reflux reactions. It is noted that corrosion problems require all components to be made of "appropriate plastic materials" [48d]. It is also emphasized that the expensive process of converting uranium oxide to uranium hexafluoride is unnecessary in the Chemex process [48d].

The American process

The characterization of this process as 'American' is a bit arbitrary [23c], since it is clear that it has been studied by researchers in many countries. It is also clear that this process is not nearly as close to commercial realization as the Japanese and French processes mentioned previously. Therefore only a brief summary of its current status will be given here.

The process involves the exchange of UF_6 molecules in the following reaction:

$$NO^{238}UF_6 + {}^{235}UF_6 \rightleftharpoons NO^{235}UF_6 + {}^{238}UF_6$$

The most extensive experiments have been done with two liquid phases: $NOUF_6$ dissolved in anhydrous hydrogen fluoride (HF) and the UF_6 dissolved in freon or some other saturated fluorocarbon. The measured equilibrium constant for this reaction is between 1.0010 and 1.0016 [27d] and the equilibrium time seems to be comparable to that of the French process, that is, of the order of one minute [15d]. These data suggest that the process might be potentially attractive, but all efforts to develop it past the laboratory stage seem to have encountered severe difficulties with the reflux reactions.

The major problem seems to be the reaction which recycles $NOUF_6$ [27d, 15e]. No way has yet been found to accomplish this reflux in a reasonable time and without excessive energy costs and losses of material. It is not clear from the open literature whether or not these problems are still being studied, but in any event this process appears to be far behind the other two discussed above. The process does not seem to have a high priority in US research and development plans [49a].

Summary

This survey of chemical enrichment methods has shown that at least two processes have evolved to the point where they have begun to look commercially attractive. Supporters of both the Japanese and the French processes point out the advantages of the relative simplicity and low energy costs of the technology, and also emphasize that the process lends itself to the economical construction and operation of small or moderately sized units [38d, 46c]. But at the same time they argue that the large inventories, long equilibrium times and criticality dangers (the danger of an accidental chain reaction) all operate to prevent the misuse of such small facilities for the production of highly enriched uranium [38d, 50a]. These claims are examined in chapters 2 and 3.

VI. Laser isotope separation

Basic principles

Just as in chemical exchange, laser isotope separation depends on the quantum mechanical connection between energy and frequency in a molecule or atom. In chemical exchange this dependence is quite subtle and indirect, but, in contrast, the use of lasers to separate isotopes exploits this connection in an elegantly simple and direct way. This is made possible by the unique properties of the laser.

According to the quantum theory an atom or molecule cannot have an arbitrary energy but can exist only in a set of discrete 'states', each having a well-defined energy. The number of such states can be very large, but they are all separated from each other by finite energy differences. The atom or molecule can be induced to make transitions from one such state to another by the absorption or emission of electromagnetic radiation. This radiation is in the form of a 'photon', an entity which exhibits either particle-like or wave-like properties, depending on the way it is observed. Among its wave-like properties is a frequency of vibration, v, related to its energy by the simple equation

$$E = hv$$

where h is Planck's constant equal to 6.63×10^{-34} joule seconds (J s).

A laser is a device which can produce large numbers of such photons, all having almost precisely the same frequency. If this radiation can be focused on a gas of atoms or molecules which have a pair of states differing in energy by just this amount, then the atoms or molecules can absorb the laser light and be 'excited' to the higher energy state. The most important characteristic of this excitation process is that it can only occur with large probability if the photon energy is 'tuned' precisely to the energy difference of the atomic or molecular transition. In this way the process resembles the operation of an ordinary radio or television set, which can be tuned to receive and amplify only one particular frequency out of the many which simultaneously impinge on the antenna.

In an atom the energy differences between states depend on the detailed structure of the electron cloud as well as on the properties of the nucleus. As was pointed out in chapter 1 the structure of the electron cloud is determined almost entirely by the number of protons in the nucleus, but changing the number of neutrons does produce small effects. For uranium, typical shifts in the absorption frequencies are only about 1 part in 100 000. For example, the energy of an important transition for isotope enrichment of uranium is about 2.1 eV,[3] and the difference between the energies for this transition for ^{235}U and ^{238}U is 4.2×10^{-5} eV [51a]. This means that a laser must be tunable to an accuracy of 1 part of 10^5 in order to excite one of the isotopes without affecting the other. Such fine tuning is possible with modern lasers. In particular a class of lasers called dye lasers operate in the visible part of the electromagnetic spectrum using organic dyes. These dyes can be designed to produce radiation of any desired colour, and within a given colour they can be tuned to very narrow frequency bands [2bb]. For example, the light which induces the above-mentioned transition is red-orange in colour and produced from a very common dye called rhodamine 6G [52].

If molecules are used instead of atoms, the transitions involve changes in vibrational energy of the molecule rather than electron energy states. Generally vibrational transitions require considerably less energy than electronic transitions, so the laser light required is in the infra-red rather than the visible portion of the spectrum. An important example is a vibrational transition in UF_6 which has an energy of 0.065 eV, quite far out in the infra-red [2cc].

Powerful infra-red lasers which use carbon monoxide or carbon dioxide are available at a number of frequencies, but it has not been as easy to produce this particular frequency as it was in the atomic case. There is some disagreement in the literature about how close this problem is to solution [49b, 42b]. Meanwhile some researchers have taken another track and instead of attempting to bring the laser to the molecule they are attempting to design a molecule to fit the laser. The most promising laser is the reliable and powerful carbon dioxide laser, which operates at a wavelength of about 10 μm (photon energy approximately 0.1 eV). A number of attempts have been made to design molecules containing uranium which will have strong vibrational transitions in this energy range [53, 54]. This is not unlike the problem in chemical exchange of trying to design uranium molecules or complexes with large chemical isotope effects.

Currently the development of laser enrichment is proceeding along several parallel paths. For the more detailed descriptions which follow it is

---

[3] The symbol eV stands for electron volt, an energy unit convenient for atomic and molecular processes. One electron volt equals 1.6×10^{-19} J.

convenient to divide the field into three categories: those which use atomic uranium, those which use molecular species, in particular UF_6, and a newer and much less developed class which might be called laser-assisted aerodynamic or diffusion processes.

Atomic vapour laser isotope separation (AVLIS)

Research and development on the AVLIS process is under way in many countries, but the system on which the most information is publicly available is the one which was until recently undergoing development by Jersey Nuclear–AVCO Isotopes (JNAI) in the USA. This project was terminated in March 1981 because of a substantial cut in financial support by the US Department of Energy which decided to direct most of its support to a similar project at the Lawrence Livermore Laboratories (LLL) [55]. The reasons offered for the decision were technical, but given the history of competition between the two laboratories and the fact that the decision directly overruled the recommendations of a technical advisory panel [49c], one cannot escape the suspicion that the reasons were at least as political as they were technical.[4]

Despite the abandonment of the JNAI programme our description will be based largely on the JNAI technique. The rather small amount of information available on the LLL process strongly suggests that it is quite similar to the JNAI process, although the Livermore group seems to be keeping open a number of alternative options for laser wavelengths and pumping schemes [56]. However, a description based on the JNAI process will certainly be adequate for the purposes of this survey. It should also be noted that research on laser enrichment is being actively pursued in a number of other countries, including, among others, the USSR and Israel. There have also been some reports of an Indian laser research programme (see chapter 8). However, very little information is available on these programmes.

An AVLIS module

The basic features of an AVLIS module are illustrated in figure 6.18 [51b]. Such a module might measure about one metre in height and one to three metres in length, and it would be entirely enclosed in a system capable of

---

[4] As this book was going to print, the US Department of Energy announced that it had chosen the Lawrence Livermore AVLIS process over its two competitors, the Los Alamos MLIS process and the TRW Corporation plasma process, for further development [105]. The Livermore AVLIS process will now be the only advanced enrichment method brought to the pilot plant stage in the USA. It is also being explored for possible use in isotopically purifying plutonium for the US nuclear weapon programme (see below, p. 175 and ref. [82]).

Figure 6.18. An AVLIS module
Uranium vapour is produced by electron beam evaporation from the crucible at the bottom. The vapour expands outward until it reaches the irradiation region where the ^{235}U atoms are excited and ionized by laser light. The ions are deflected and collected on the vertical plates by a combination of electric and magnetic fields. Neutral atoms continue outward and are collected on the horizontal plate at the top. The laser light is shown being reflected several times through each collection volume by a system of mirrors.

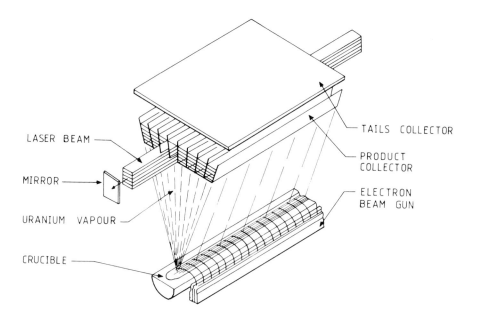

maintaining a high vacuum. All laser light sources and mirrors would be outside the volume containing uranium vapour. Since the path length of the laser light in the vapour may be more than 100 m [57a], the light must be reflected back through the chamber many times, possibly as many as 300 times [42c].

The uranium to be enriched is first converted to the pure metallic form and formed into a long ingot. It is then melted in a crucible by heating it with a beam of electrons directed to the surface of the ingot by a magnetic field. Since hot liquid uranium is extremely corrosive, a rather elaborate mechanism must be used to keep the molten uranium from contacting the support structure and to prevent rapid dissipation of the heat supplied to the uranium ingot [2dd]. The thin strip of molten uranium provides a line source of uranium atoms which then diverge radially outwards towards the top of the chamber. The atoms are allowed to move undisturbed through the lower portion of the chamber during which time they lose much of the excess energy given to them in the evaporation process. The purpose of this is to get as many atoms as possible into their lowest energy states and to allow atoms ionized by the electron beam to recombine into neutral

161

form. Creation of this radially diverging vapour involves considerable losses, and only 50 per cent of the evaporated atoms reach the irradiation zone [42d]. The rest are deposited on various surfaces inside the chamber, and this material must be periodically collected and recycled. It represents a kind of internal reflux analogous in principle to the recycled intermediate fraction in the advanced Becker nozzle process (see p. 149). However, in contrast to the efficiency improvements this reflux leads to in the nozzle process, in the AVLIS process it represents a significant waste of both energy and time.

The vapour that does reach the irradiation zone is then illuminated by laser light carefully tuned to excite transitions only in ^{235}U atoms. Four lasers are used, all with slightly different colours, but all in the same red-orange portion of the spectrum. The need for four lasers can be understood by referring to figure 6.19 [51c].

The total energy required to remove an electron from a uranium atom is 6.2 eV. This is very difficult to supply with a single laser in an isotopically

Figure 6.19. Three-step laser ionization
The first step uses two laser beams of slightly different frequency in order to excite the large number of ^{235}U atoms in the ground state and low-lying excited state. The ^{238}U level scheme is similar to this one for ^{235}U except that each level is shifted by an amount too small to be shown on the figure, but large enough to allow discrimination by finely tuned lasers.

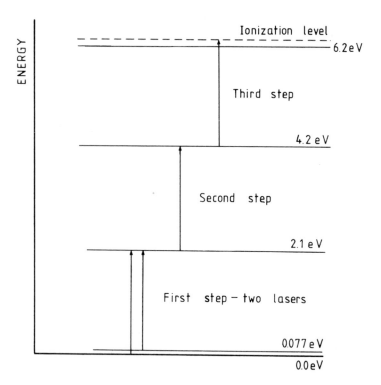

selective way, although the argon fluoride laser may turn out to be useful for this purpose [56a]. In practice, at least two transitions are required, and three have been selected in order to take advantage of the readily accessible and well-established technology of rhodamine dye lasers. Notice that the ionization energy of uranium is just under three times the energy of a red-orange photon, so three steps is just the right amount. Each of these steps has a slightly different energy, so the three lasers must be tuned to slightly different frequencies. These three lasers are sufficient to ionize about 40 per cent of the ^{235}U atoms in the vapour. This is the fraction which is in the lowest energy state at the high temperature at which the irradiation takes place. Another 30 per cent are in the next higher energy level, which is only 0.077 eV above the ground state. It turns out to be economically advantageous to employ a fourth laser to get at this additional 30 per cent. Even so, the remaining 30 per cent of the ^{235}U is unaffected by the lasers and becomes part of the tails.

The ^{235}U atoms which are ionized by the laser light are then deflected by a pulse of strong electric and magnetic fields towards collecting plates oriented parallel to the radial flow of neutral vapour. Ideally only the charged ions would be given transverse velocities and migrate to the collector plates, and the neutral atoms would continue to move outward and be deposited on a surface beyond the irradiation-collection region. In practice, however, it is not possible to shield the collectors from all of the neutral particles, and anywhere from 3 to 15 per cent of the feed material is expected to be collected along with the enriched material. This and several other inefficiencies limit the achievable single-stage enrichment factor to 15 at the very most [42e].

If such a value can be obtained in practical enrichment facilities, it will constitute a remarkable advance over current processes, for which enrichment factors are only slightly larger than one. For example, such a facility would be capable of producing 3 per cent reactor fuel from 0.2 per cent feed in one stage, thereby creating an enormous source of fuel from the existing stockpile of gaseous diffusion plant tailings. This possibility seems to be one of the major driving forces behind the development of laser enrichment. It must be emphasized, however, that an enrichment factor of 15 is a highly optimistic goal. It is probably more reasonable to expect a value between 5 and 10.

These numbers can now be used to construct a model of an AVLIS enrichment stage (figure 6.20). The numbers on the model were derived assuming natural uranium feed, an enrichment factor of 7, and a reflux to feed ratio of 0.5. It was also assumed that 60 per cent of the ^{235}U which makes it to the irradiation region is ionized and collected. Amounts of feed, reflux and tails are computed on the basis of 1 kg of product. The amount of separative work done by this stage can be computed from equation 5.4 to be 7.6 kg SWU/kg product. Note that for such large enrichment factors the use of quantities such as cut and separation gain is not meaningful. For this reason it is the value of the enrichment factor, that

Figure 6.20. An AVLIS enrichment stage

All flow numbers are based on 1 kg of product and a hypothetical enrichment factor of 7. The reflux represents the 50 per cent of feed material which deposits on various surfaces of the module and is unprocessed. This material must be recovered and recycled.

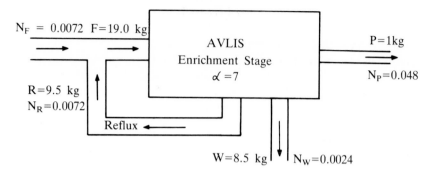

is, the ratio of isotopic compositions of product to feed (see equation 5.7, p. 104) which is entered in table 6.3.

Another standard parameter which loses much of its significance is the stage hold-up time. If this is taken to be the time required for a uranium atom to move from the surface of the liquid to the collector, it is only a small fraction of a second. Obviously this is not a controlling factor in determining production rates or equilibrium time. The latter is essentially instantaneous. Far more important are the materials-handling processes which produce feed and recover product and tails. These require highly elaborate and sophisticated procedures [51d], which present important and apparently still unresolved problems [49d].

One of the major early attractions of the laser enrichment processes was their potential for great savings in energy over gaseous diffusion and even the centrifuge [58]. The specific energy consumption of the element of figure 6.20 can be estimated by assuming that 30 per cent of the ^{235}U atoms in the feed absorb 6.2 eV of photon energy to produce 7.6 kg SWU. This represents 1.0×10^5 J or 2.8×10^{-2} kWh. Even if the lasers are only 0.2 per cent efficient, the energy required is only 14 kWh or about 2 kWh/kg SWU. This is 1000 times less than gaseous diffusion and 100 times less than the centrifuge. However, this estimate ignores a number of other factors, such as evaporation energy, optical system inefficiencies, reflux procedures, and so on. More recent estimates of energy consumption make AVLIS comparable to or somewhat better than the most efficient centrifuges, that is, about 100 kWh/kg SWU [2ee, 56b].

Laser requirements

Energy costs, however, are only a part of the problem, and when it comes to capital costs and technical complexity in design and operation the laser

164

process looks much more problematical. As only one example of the extreme complexity of the system it is instructive to examine the requirements which the lasers must satisfy.

First, the lasers must be able to provide a certain minimum flow of photons which is determined entirely by the properties of the atomic transition itself and is independent of the density of vapour or the desired production rate. This requirement follows from the need to make sure that sufficient numbers of atoms are available at each of the energy steps to take advantage of the photons ready to boost them to the next step. Because excited atomic states decay, it is essential that they be re-excited as fast as they decay, and this puts a lower limit on the photon flux which is acceptable. In laser jargon this requirement is that the energy level be 'saturated' [59a] and this saturation requirement means that the lasers must be quite powerful, even to produce relatively small quantities of product. The minimum power density required can be shown to be of the order of tens of kilowatts per square centimetre. This can be produced by a laser-pulsed beam of 1 cm^2 containing tens of millijoules of energy and lasting one millionth of a second.

Another stringent requirement placed on the lasers is the pulse repetition rate. As the uranium vapour passes through the irradiation zone it is crucial that all the ^{235}U atoms be sufficiently exposed to the laser radiation. This means that two pulses of radiation cannot be separated by a time longer than it takes a uranium atom to cross this zone. If the zone is assumed to be 5 cm thick, and the average speed of a uranium atom is 500 m/s, then it takes only 10^{-4} s for an atom to cross the zone. So the lasers must be fired at least 10 000 times per second to ensure full exposure of the uranium. Since it is highly unlikely that a single laser with sufficient beam intensity could be made to fire at such a rapid rate, the AVLIS system might require 20 or more sets of four lasers each, all precisely controlled in frequency by a master laser oscillator and timed to fire in sequence, each one at a rate of 500 pulses per second [51e]. It is probably unnecessary to emphasize that such a system presents major technical problems, both in its original design and in ensuring that it can work for hours at a time in reliable, continuous operation.

One issue which has generated some controversy is the ease with which an AVLIS enrichment stage could be used for batch recycling to obtain highly enriched product. With an enrichment factor of 10, only three such recyclings would be needed in principle to produce 97 per cent enriched product. However, one evaluation of the proliferation dangers of this technique has argued that such a procedure would be extremely difficult, if not impossible [42f]. Briefly stated, the argument is as follows:

If normal uranium vapour densities are used, then as a larger fraction of the uranium becomes ^{235}U, the laser power must be increased in order to ionize the same fraction of ^{235}U atoms, or, if the ^{235}U content is over 50 per cent, the lasers can be tuned to remove ^{238}U instead. If laser powers are increased, then the ^{235}U plasma created by the ionization becomes so dense that the efficiency of the ion collectors drops. This is a plasma shielding effect similar to those discussed in the next section (see pp. 180–81). Tuning the lasers to excite ^{238}U will work only if the plasma shielding effects remain small, up to 50 per cent ionization. Finally, if the density of the atomic vapour is reduced in an attempt to solve the problem, there will no longer be enough collisions in the vapour to bring a sufficient number of atoms into their lowest states. Again the efficiency of the process will be degraded.

These arguments are all physically valid, but a question remains as to their quantitative significance. Some knowledgeable analysts seem unconvinced by these arguments and suggest that the batch recycling procedure is feasible as long as some sacrifice in throughput is tolerable [60,61]. It is not possible to make an independent judgement on this question using only open sources of information.

Whatever the merits of this particular argument, it can be safely said that the creation of a successful AVLIS facility will be an extremely complex and sophisticated technical achievement. The laser system will involve hundreds of powerful lasers all pulsed according to a precisely designed time schedule. The light from the lasers will have to be reflected hundreds of times through a system involving hundreds of extremely efficient mirrors and other optical devices, all of which must be shielded from any contact with the uranium vapour. The optical system is further complicated by the need to prevent the propagation through the system of frequencies corresponding to other uranium transitions. If these are allowed to pass through the system, then the uranium vapour can itself become a laser and lose most of the energy pumped into it without even being ionized. Methods exist for dealing with this problem, but they add considerably to the complexity of the laser optical system [62].

There are even more subtleties and difficulties which could be mentioned, but the above should suffice to demonstrate that the AVLIS process is enormously complex and nowhere near commercial viability. This has been dramatized most vividly by the recent withdrawal of the JNAI effort from the competition.

Molecular laser isotope separation (MLIS)

Although some efforts have been made to use molecules better suited to existing lasers (see p. 159), the major efforts to develop MLIS have concentrated on UF_6. The reasons for this are obvious: UF_6 is a well-understood material with a high volatility and convenient chemistry. It is not an easy substance to handle in practice, but so much experience has been gained in its management from the gaseous diffusion and centrifuge programmes that it presents no serious difficulties.

The basic principle of the MLIS process is similar to that of AVLIS. $^{235}UF_6$ molecules must be made to absorb energy from laser beams while

$^{238}UF_6$ molecules remain unaffected. However, the different nature of the absorption process makes the use of dye lasers inappropriate, and MLIS relies instead on either all infra-red light or a combination of infra-red and ultraviolet. The first process is often called 'infra-red multiphoton dissociation' and the latter 'two-step dissociation'. In each case the object is to inject enough energy into a $^{235}UF_6$ molecule so that it loses one of its fluorine atoms:

$$^{235}UF_6 \xrightarrow{\text{energy}} {}^{235}UF_5 + F$$

The $^{235}UF_5$ which is produced condenses rapidly into a fine powder which can then be filtered from the UF_6 gas [49e].

In the AVLIS process the use of hot uranium vapour reduces the efficiency of the process since only a minority of the uranium atoms can be excited out of the lowest energy state. This problem requires an extra laser and still results in a loss of 30 per cent of the potential ^{235}U atoms. This same problem occurs in the molecular process, and in fact is far worse. At ordinary temperatures the collisions between UF_6 molecules are so violent that virtually all of the molecules are excited out of their lowest vibrational states. The molecules have a wide range of vibrational energies which makes it very difficult to get any significant selectivity by tuning to a particular vibrational transition. Only by cooling the UF_6 to very low temperatures can this problem be solved. But at very low temperatures UF_6 is normally a solid.

The solution to this problem is to supercool the UF_6, that is to cool it to temperatures where under normal conditions it would condense, but to trick it into believing it is still a gas. This is done by diluting the UF_6 with an inert carrier gas such as nitrogen or argon and expanding the mixture suddenly through a nozzle (see figure 6.21). The expanding gas cools rapidly, and as the molecules collide downstream of the nozzle the vibrational energy in the UF_6 is converted into the translational motion of the gas [57b]. This is quite similar to the process which occurs in the atomic vapour between the evaporation point and the irradiation region. However, the molecular process is more successful, and roughly 95 per cent of the UF_6 molecules can be put into the lowest vibrational state. Therefore, only a single isotopically selective laser is needed to excite most of the $^{235}UF_6$.

The wavelength of the light required from this first laser is 16 μm, and a number of possible candidates already exist. One is a combination carbon dioxide (CO_2) and carbon tetrafluoride (CF_4) system developed by a group at the University of Southern California [63]. Another uses Raman scattering in hydrogen to step up the wavelength of CO_2 laser light from 10 μm to 16 μm [64, 65, 66].

A third potential candidate is the recently developed free-electron laser, a remarkable new kind of light source which shows promise of both high efficiency and precise tunability [67]. There are many incentives aside from uranium enrichment for the development of these lasers, so it is reasonable to assume that one or more of them will be perfected.

Figure 6.21. An MLIS stage

This is a schematic illustration of the components of an MLIS stage. UF_6 and the inert carrier gas are mixed and expanded at supersonic speeds through a nozzle. Just downstream of the nozzle the gas is irradiated by an isotopically selective infra-red laser which vibrationally excites the $^{235}UF_6$, and a powerful ultraviolet laser which dissociates the excited UF_6 molecules. The $^{235}UF_5$ 'laser snow' is filtered from the gas stream and sent on to be refluorinated back to UF_6. The remaining gas is cleaned up and sent on for further tails stripping.

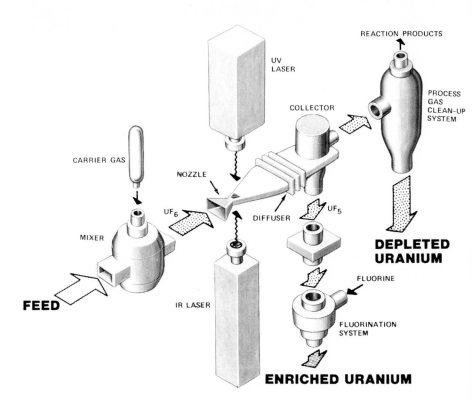

Source: Dr N. Haberman, US Department of Energy, Washington, D.C., USA.

When the $^{235}UF_6$ molecules have absorbed a 16 μm photon they can then be further excited either by the absorption of many more infra-red photons, or by a single ultraviolet photon which carries enough energy to dissociate the molecule directly (see figure 6.22). The tendency of complex molecules to absorb a number of infra-red photons of the same frequency was first noticed in 1974 [68] and has since been a topic of intense experimental interest [59b]. One of the strong advantages of this method seems to be that once the molecule has absorbed a small number of 16 μm photons the rest of the infra-red absorption can be supplied by the powerful and efficient CO_2 laser at 10 μm [69]. On the other hand the multiphoton absorption process is still not very well understood, and

Figure 6.22. Two molecular dissociation processes

The horizontal lines represent vibrational excitation levels, and the vertical arrows transitions induced by laser radiation. In the scheme on the left a large number of infra-red photons are used to excite the molecule up through many vibrational levels to dissociation. In the process on the right only one infra-red photon is used to excite $^{235}UF_6$, and the dissociation is accomplished in one step with an ultraviolet photon.

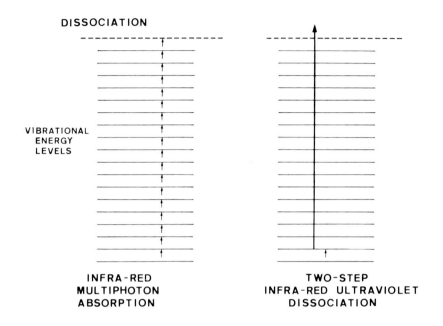

DISSOCIATION

VIBRATIONAL
ENERGY
LEVELS

INFRA-RED	TWO-STEP
MULTIPHOTON	INFRA-RED ULTRAVIOLET
ABSORPTION	DISSOCIATION

certain features of it suggest that it may be difficult to apply the process to the large quantities of UF_6 required in a commercial-sized enrichment plant [70]. One particularly troublesome phenomenon is the tendency of irregularities in laser beam intensity to become magnified by a non-linear process called self-focusing. This can cause the beam to shrink down to very small irradiation volumes or to break up into many 'filaments' [71, 72]. The recent discovery of this phenomenon calls into question the validity of previous interpretations of infra-red absorption experiments and raises questions about the applicability of the process to large-scale processing of UF_6 [72a].

The alternative of using an ultraviolet laser for the second step is the method currently being pursued most actively by the Los Alamos Laboratory in the USA. There seems to be some doubt that this process may be successfully developed for large-scale enrichment. The basic problem is the very low absorption probability for ultraviolet photons in vibrationally excited UF_6 molecules. This implies that even under favourable conditions the UV excitation process will be very inefficient, and that the UV lasers will have to be large and powerful. The open literature suggests a genuine dispute among researchers concerning the prospects for UV photo-dissociation. Even a relatively optimistic assessment points out that "the . . . UV laser system . . . will require substantial development to achieve low-cost, reliable devices" [49e]. Others are quite negative, even to the point of excluding ultraviolet photo-dissociation from a list of promising processes for collecting vibrationally excited UF_6 molecules [73]. There is no way for an outsider to decide who is right in this argument, but it is safe to say that arguments of such a fundamental nature generally do not take place over systems which are close to commercial application.

As difficult as the laser problem may be, the advantages of the MLIS process, as well as the many other potential uses for such lasers, guarantee that the search for adequate lasers will go on. The MLIS process promises much smaller irradiation volumes than AVLIS and therefore less extensive and complex optical systems for the laser beams. This follows from the higher densities possible in the irradiated gas [57c]. There is also the advantage of handling UF_6 instead of hot, corrosive uranium vapour. Finally, there is the possibility that the molecular process can be conveniently operated in stages. This follows from the simple and clean method by which the enriched product in the form of UF_5 'laser snow' can be filtered from the gaseous tails stream [74]. The latter can immediately be sent on for further depletion while the former can be quickly refluorinated (see figure 6.21) and sent on for further enrichment.

The actual feasibility of this will depend on how sensitive the process is to ^{235}U assay. However, there is no obvious analogy with the problems encountered for higher assays in the AVLIS process (see pp. 171–72). In that process there was no buffer gas to dilute the ^{235}U concentration to match the laser power capabilities while still promoting vibrational cooling. So there seem to be no obvious obstacles to cascading the MLIS process to high enrichments. This means that one solution to the ultraviolet laser problem may be the use of many lower power lasers in a cascaded system rather than a single very powerful UV laser in a one-stage system. It is perhaps significant that the most recent theoretical work on non-ideal cascades with high single-stage enrichment factors has been done at Los Alamos [75, 76].

One other important difference between the MLIS and AVLIS processes is in the product collection mechanism. Not only does the formation of UF_5 snow facilitate the segregation of product from tails, but it also removes the need for very high laser pulse rates. These are needed in the AVLIS process to ensure irradiation of all of the vapour, since non-irradiated vapour is collected at a fixed rate along with the ^{235}U ions. But in the MLIS process any UF_6 gas which is not irradiated simply continues on into the tails stream and has no effect on the product assay of the irradiated vapour. In addition it would seem that for higher feed assays less degradation of product would occur as a result of fluorine-exchange reactions. This exchange process

$$^{235}\text{UF}_5 + {}^{238}\text{UF}_6 \rightleftharpoons {}^{235}\text{UF}_6 + {}^{238}\text{UF}_5$$

can take place when the photo-dissociated $^{235}\text{UF}_5$ collides with $^{238}\text{UF}_6$ molecules before condensation. Such collisions are of course less frequent the higher the percentage of $^{235}\text{UF}_6$ in the gas. So one expects that enrichment efficiency would improve at higher assays, again adding to the attraction of the cascading. The use of a buffer gas for vibrational cooling means that the $^{235}\text{UF}_6$ density can easily be made independent of feed assay and thereby always matched to the capabilities of the lasers.

No data are available, so the best that can be done is to estimate that the separation factors achievable in the MLIS process will be comparable to those of the AVLIS process. If not, the greater ease of cascading should redress the balance. Energy consumption in the laser systems of both processes is roughly equivalent, and the evaporation energy of AVLIS is replaced by pumping energy in MLIS, so it seems reasonable to assume that the MLIS process will share the energy-saving virtues of AVLIS.

All of these factors suggest that, if the laser problems can be solved, the MLIS process may actually prove to be effective for producing highly enriched uranium at significant rates in relatively small facilities. This has clear implications for proliferation control, which are examined in chapter 2.

Laser-assisted processes

The foundation of most laser-assisted processes is the ability of the laser to selectively deposit energy in one kind of molecule while leaving all others unaffected. In the previously discussed methods this energy is used to ionize an atom or dissociate a molecule, but even if smaller amounts of energy can be deposited selectively, some very interesting results can occur. In particular, this energy can be used to rapidly raise the temperature of one component of a gas mixture while leaving the temperature of other components relatively unaffected.

The concept of thermal equilibrium was discussed in section II in connection with the theory of gaseous diffusion (see p. 121). Any dense mixture of gases will rapidly achieve thermal equilibrium and an equipartition of energy by virtue of the many collisions between molecules. This equipartition of energy fixes the ratios of average velocities of the molecular species, and in the case of $^{235}\text{UF}_6$ and $^{238}\text{UF}_6$, this ratio is only very slightly different from one. This very slight difference then leads to the requirement of many enrichment stages in a gaseous diffusion or aerodynamic separation facility.

Suppose, however, that it were possible to raise the temperature of the $^{235}\text{UF}_6$ relative to the $^{238}\text{UF}_6$. This average velocity ratio could then be substantially increased and far fewer stages might be needed. This selective increase in temperature can be achieved with an infra-red laser similar to the one used in the MLIS process. The laser excites $^{235}UF_6$ molecules into a high vibrational state, and, then, as the molecules collide with others, usually a buffer gas, much of this vibrational energy can be converted into translational energy, that is, higher speed. The $^{235}UF_6$ will then diffuse much more rapidly than $^{238}UF_6$.

One proposal to this effect has been made by JNAI [77]. The process involves injecting a subsonic flow of a UF_6–argon mixture into a tube in which it can be irradiated with an infra-red laser. At the downstream end of the tube are a set of cryogenically cooled plates through which all the gas must pass. Since the $^{235}UF_6$ molecules have been given higher velocities, they are more likely to collide with the plates where they condense to form solid UF_6. After some period of operation the tube is evacuated and the plates are heated to release the accumulated UF_6 product. This can then be sent directly on to another stage without any reflux chemistry having to be performed.

Another suggestion has been to combine a laser with the Becker nozzle process or other aerodynamic processes [78]. Since the selectivity of these processes depends on the rate of diffusion of $^{235}UF_6$ across curved streamlines, the separation effect can be enhanced by selectively heating the $^{235}UF_6$ as it passes through the curved region, thus enhancing its rate of diffusion.

A third proposal uses atoms of uranium instead of UF_6 and proposes that the laser photons be used directly to impart a transverse momentum to the ^{235}U atoms in a collimated atomic beam [79]. When an atom absorbs a photon it acquires that photon's momentum; so if the beam is not too dense and an adequate collision-free drift region can be maintained, it may be possible to separate the deflected beam from the undeflected one and obtain an enriched product.

No evidence exists that these methods are much more than interesting ideas at this time, and their ultimate fate will take a long time to be determined. Meanwhile new variations on this basic theme and others are appearing with undiminishing regularity. The number of possible variations seems to be very large.

Summary

This survey has shown that a large number of isotope separation techniques have been made possible by the advent of the laser. The essence of the laser's attractiveness for this purpose lies in its highly precise ability to process only ^{235}U, leaving the vast vast majority of the uranium unaffected. This increases selectivity and greatly reduces energy costs. However, it achieves these improvements at the cost of high technical sophistication and component costs. Substantial questions still remain as to whether or not stable, reliable, and economical laser and materials handling systems can be developed.

If they are developed the implications for weapon proliferation of laser isotope separation are quite serious. Attention was called to these implications in 1977 [57, 80, 81], but since that time there has still been relatively little public discussion of the costs and benefits of continued development of these systems. As usual, concern about proliferation seems to be subordinated to commercial and bureaucratic interests.

One more important aspect of laser processes deserves mention. This is the ability of the laser to also separate plutonium isotopes, something which is not generally considered feasible in any of the other processes discussed previously [82]. The most important plutonium isotope for weapons is ^{239}Pu, but this tends to be heavily contaminated by ^{240}Pu when the plutonium has been produced in normal nuclear reactor operations. The presence of the ^{240}Pu degrades but does not destroy the material's usefulness as a nuclear explosive [50b].

Since the two isotopes differ in mass by only one unit instead of three (as in uranium), and since other isotopes (^{238}Pu, ^{241}Pu, ^{242}Pu) are also often present, the ability of all the previously described methods to separate them is greatly reduced. In addition, plutonium presents much more severe problems of radioactivity, toxicity and criticality than uranium.

Only the laser, and possibly the electromagnetic methods discussed in the next section, show some promise of separating plutonium isotopes. Presumably plutonium metal can be vaporized like uranium metal, but criticality dangers would require that the amounts used be much smaller. Plutonium also forms PuF_6, a substance which seems to have properties very similar to UF_6 [1c]. Research on laser plutonium separation is going on in the USA, and presumably elsewhere as well [42f, 49f, 83]. Given these developments there can be no doubt that continued progress in laser isotope separation will greatly complicate efforts to control nuclear weapon proliferation.

VII. Electromagnetic and plasma processes

Basic principles

All of the methods in this category depend on the ionization of all or part of the feed material and the use of electric and magnetic fields to accelerate the ions and separate the isotopes. All are characterized by relatively high separation factors but relatively low throughput, caused mainly by the inherently low densities of the ion beams or plasmas used.

A feature common to all electromagnetic processes is the use of magnetic fields to accelerate uranium ions. When a charged particle enters such a field it can experience a force, and this force has two special properties:

1. It only occurs if the particle has some motion perpendicular to the field lines.

2. The force itself is perpendicular to both the field lines and the velocity of the particle.

These two characteristics of the magnetic force lead to the result that the paths of all charged particles (electrons or ions) in a uniform magnetic field are circles or helices, with the plane of circular motion perpendicular to the magnetic field lines. The frequency of the circular motion, the so-called 'cyclotron frequency', depends only on the charge and mass of the particle and the strength of the magnetic field

$$\nu_c = eB/2\pi m \qquad (6.11)$$

Equation 6.11 shows that the frequency of rotation is independent of the particle velocity. This means that all particles complete one revolution in the same amount of time. For this to be true there must be a direct proportionality between the speed of a particle and the radius of its orbit, called the 'radius of gyration'. So, for a given type of particle the faster it is moving the larger the circle it will describe. This is the physical basis for the ion cyclotron resonance technique for isotope separation which will be described more fully below.

Another way of utilizing the magnetic force is to give all the ions a very well-defined velocity perpendicular to the field. Then the magnetic force will act with equal strength on all of them, but according to Newton's laws the lighter particles will experience greater accelerations than the heavier ones, implying that the lighter particles will move in circles of smaller radius. After one-half of a circular orbit, the beam will have separated into a number of distinct beams, each containing only particles of a given mass (assuming that all have the same charge). This is the principle underlying the technique of mass spectroscopy, employed for isotope separation in a device called the calutron.

An even more interesting phenomenon can be observed if an electric field is added to the magnetic field. For example, consider a situation such as shown in figure 6.23 in which a radially directed electric field has been superimposed on the axial magnetic field of a solenoid. The electric field will accelerate positive ions outward, but as the ions move outward their interaction with the magnetic field will cause them to deflect into circular orbits. After a very short time the ions will acquire a uniform speed called a drift velocity in the azimuthal direction, that is, perpendicular to both the electric and magnetic fields. This motion differs from cyclotron motion because the direction of the drift velocity is independent of the charge of the particle.

The magnitude of the drift velocity is also independent of both the mass and the charge of the particle, but the drift motion is superimposed on the cyclotron motion. This results in a net azimuthal velocity which does depend on the mass of the particle. If one now injects into the solenoid a

Figure 6.23. Motion of charged particles in crossed electric and magnetic fields
The magnetic field (B) inside the solenoid is highly uniform and is parallel to the axis of the solenoid. An electric field (E) is shown pointing radially outward from the axis to the periphery of the solenoid. In such a field configuration particles of both positive and negative charge will rotate anti-clockwise as shown. Their drift velocity, V_D, is determined entirely by the values of E and B.

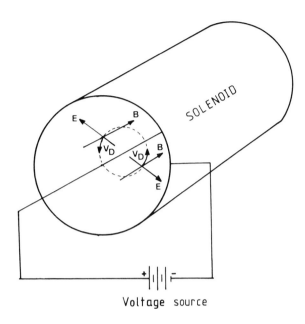

Voltage source

gas of ions and electrons, that is, a plasma, the entire gas will be set into rotation. Very high rotational speeds can be achieved in this way. This is the principle underlying several suggested designs for 'plasma centrifuges' which separate isotopes in a manner very similar to the mechanical centrifuge of section III but with no moving parts.

Before proceeding to more detailed discussions of particular processes, it is useful to describe the general features of a plasma. A plasma is a gas of charged particles — generally positive ions and negative electrons. It must have a very high temperature, at least of the order of thousands of degrees Celsius, so that the high relative velocities of ions and electrons will prevent their rapid recombination to neutral atoms. It is possible to apply many of the same laws of statistical mechanics which apply to neutral gases to plasmas, in particular the equipartition law. This immediately leads to the prediction that the electrons in a plasma must have much greater average velocities than the ions. In a uranium plasma this difference is particularly dramatic since the mass of a uranium ion is over 400 000 times that of an electron. This implies that electron velocities are about 650 times greater than ion velocities on the average.

This enormous disparity in velocities suggests that electrons might quickly escape from the plasma, but electrons which do emerge from the plasma are attracted back by the residual positive charge of the ions left behind. These electrical forces are very strong and tend to preserve the overall charge neutrality of the plasma. The result is that the plasma is surrounded by a thin sheath of electrons held in place by the attraction of the positive charge in the interior. The thickness of this sheath can be estimated from simple electrostatic principles, and it is found to be comparable to the Debye shielding distance [84]. Strictly speaking, a plasma is defined to be a gas of charged particles whose dimensions are large compared to this distance. In the uranium plasmas used for isotope separation the Debye shielding distance is typically $1-10$ μm.

Plasmas can be produced in a number of ways. One method is to inject a beam of high energy ions or electrons into a neutral background gas. Collisions cause the ionization of the background gas, and a plasma can be formed. Another method is to create an electrical discharge in a gas between two electrodes by applying a voltage difference. This is a phenomenon observed in nature: in electrical storms lightning discharges create temporary plasmas along the path of the discharge, and the strong electric fields present in the humid air can cause other strange effects which depend on the formation of plasmas, one of which is called 'ball lightning' [85].

A third method of obtaining a plasma is to allow a neutral beam of atoms to strike a hot, glowing metallic surface, which strips the electrons off the incoming atoms and forms a plasma. Such a device is called a 'Q-machine' [2ff].

With these general introductory comments as background, it is now possible to consider the three classes of electromagnetic methods in more detail.

The calutron

The name 'calutron' says a great deal about the historical origins of this process. The calutron was invented in the laboratory of Ernest O. Lawrence of the University of California (hence 'cal') in the early 1940s. In the late 1930s Lawrence had invented the cyclotron, a device for accelerating charged particles to high energies for nuclear physics research, and the calutron concept was based on the same technology (hence 'tron'). In fact, with the start of World War II and the intense interest in the United States in rapidly producing atomic bombs, Lawrence turned much of his laboratory and its equipment over to the attempt to make highly enriched uranium [86a].

The major piece of equipment which made it possible for Lawrence to begin his new project rapidly was a large magnet, 4.67 m in diameter. This magnet consisted of two circular pole faces separated by a gap in which a vacuum chamber was inserted. Inside this chamber could be placed a source of uranium ions and, 180° around the circle, a collector. The source

Figure 6.24. An early two-beam calutron

One of the two-beam calutrons designated Alpha-I at the Oak Ridge facility in 1944. The unit rests on its door on a storage dolly. The covers have been removed to show the double source at the right and the two receivers at the left. Some four-beam units were also put into service before the end of the war.

Source: Oak Ridge National Lab., Tennessee, USA, and R.G. Hewlett and O.E. Anderson, Jr.,*The New World, 1939/46* (Pennsylvania State University Press, University Park, Pa., 1962).

would accelerate the ions to a velocity such that the curvature of their trajectory in the magnetic field would cause them to arrive at the detector (see figure 6.24).

The radius r of the path of an ion in a uniform magnetic field is given by

$$r = mv/eB$$

where m and e are the ion's mass and charge, respectively, v is the ion velocity, and B is the magnetic field strength. If the ions are accelerated by a voltage V then their kinetic energy will be eV and their velocities will be

$$v = \sqrt{2eV/m}$$

When these two equations are combined, the radii of curvature of the ions in the magnetic field turn out to be

$$r = (1/B) \sqrt{2Vm/e} \tag{6.12}$$

Since all quantities but m are the same for different ion species, it follows that the radius for a particular isotopic ion is proportional to the square root of its mass. This implies that two ions whose masses differ by 1.27 per cent will have radii that differ by only 0.64 per cent. Given typical values [86b]: $B = 0.34$ tesla, $V = 35\ 000$ volts, $e = 1.6 \times 10^{-19}$ coulomb, $m = 3.9 \times 10^{-25}$ kg, the beam radius comes out at $r = 1.22$ m, and the separation between the $^{235}U^+$ and $^{238}U^+$ beams would be a little over 15 mm. This is just large enough to allow a collector to be made which will admit only the ^{235}U beam and reject the ^{238}U beam with a high degree of selectivity [87a]. It even proved to be possible to put two or even four source and collector pairs into the same vacuum tank to make more efficient use of the magnetic fields (see figure 6.24).

What appears in theory to be a very simple and precise method for isotope separation turns out in practice to be extremely difficult if any more than laboratory-sized amounts of product are desired. The major problems encountered in the Y-12 calutron project were the severe limitation on ion beam strength created by space charge effects, and the inability to convert more than a small fraction of the feed material into product in any single run. Space charge problems result from the tendency of particles with similar charges to repel each other and destroy the beam. This can be overcome to some extent by allowing the positive ions to collide with residual background gas in the calutron and thereby release electrons which neutralize the repulsion. In effect, a neutral plasma is created along the beam path [11d]. However, this technique is limited to low densities of background gas, since too high a density would destroy the beam by causing too many collisions with the beam particles. So the currents in a typical calutron beam were limited to a few hundred milliamperes, leading to a collection rate of only about 100 mg/day of ^{235}U. At this rate it would take over 400 years to collect a critical mass. This explains why the Y-12 project employed more than 1 100 calutrons [87b].

The actual enrichment factors achieved in the Y-12 calutrons were at least 20 [86c] and probably somewhat over 30. Various improvements since 1945 have resulted in the capability to get enrichment factors anywhere from 30 to as high as 80 000 in a single pass [87c]. However, the very high values can only be obtained using special geometrical arrangements and very low throughput. It seems reasonable to infer that at the throughput rates needed for the production of significant amounts of highly enriched uranium a range of 20 to 40 should adequately describe the capabilities of the calutron.

The high enrichment gain is offset by a very low efficiency in feed utilization. This is due to the tendency of ionized uranium vapour to deposit on all available surfaces inside the vacuum chamber. Only 10 to 15 per cent of the feed material is actually processed [11e], meaning that the cut is below 0.01 for low-enriched feed.

This problem was particularly serious in the early calutrons which operated on feed material that had already been partially enriched and was therefore extremely valuable. The insides of these machines had to be periodically cleaned and the material recovered with high efficiency.

The low collection efficiency also leads to a very high specific energy consumption because of the high energy to which the full beam must be accelerated ($\simeq$ 35 000 eV) to get separation. Using this energy value and an enrichment factor of 30, and assuming that 15 per cent of the ^{235}U in the beam is captured, the specific energy consumption can be estimated at 3 800 kWh/SWU, greater than for gaseous diffusion. This number applies to the early calutrons, so it could possibly be improved upon with modern technology.

In the end it proved possible for over 1 000 calutrons to produce enough uranium for one bomb, but only after three years of intense effort and roughly a billion dollars of expense. It is clear that if it had not been for the overwhelming desire to achieve a usable weapon before the end of the war, the calutron project would never have been pursued as far as it was. Indeed, once the gaseous diffusion plant at Oak Ridge began to produce highly enriched product, the large calutrons were closed down. Now this process is considered suitable only for the production of small quantities of very pure isotopes. It is perhaps worth noting that the calutron method, like the laser methods, is one of the few suitable for separating plutonium isotopes [88]. But the limitations of the method make it extremely unlikely that it would ever be feasible for the production of kilogram quantities of pure ^{239}Pu from irradiated reactor fuel.

Ion cyclotron resonance

A schematic diagram showing the essential features of this process is given in figure 6.25. The method uses selective excitation of ^{235}U cyclotron motion in a plasma. The plasma is created either by an electrical discharge or by a Q-machine and caused to flow axially through a region containing a strong magnetic field created by a superconducting[5] solenoid [89a]. Typical experimental values of the field strength have been about 2 tesla [49g].

As they pass through the magnetic field the ions are subjected to an alternating electric field whose frequency is tuned precisely to the cyclotron frequency of the ^{235}U ions (see equation 6.11, p. 174). Assuming that the uranium ions are singly charged ($e = 1.6 \times 10^{-19}$ coulomb), and using an ion mass of 235 atomic mass units (1 amu = 1.66×10^{-27} kg), the cyclotron frequency of a ^{235}U ion turns out to be 130 kHz in a field of 2.0 tesla. The difference between the ^{235}U and ^{238}U frequencies is about 1.25 per cent of this, or about 1.65 kHz.

---

[5] Superconductivity is a phenomenon which occurs in many metals at very low temperatures, that is, a few degrees above absolute zero. It is characterized by zero electrical resistance and therefore a zero rate of resistive energy loss. Only liquid helium can be used to cool the magnet coils, since only helium remains a liquid at such low temperatures.

Figure 6.25. Schematic illustration of the plasma separation process based on ion cyclotron resonance

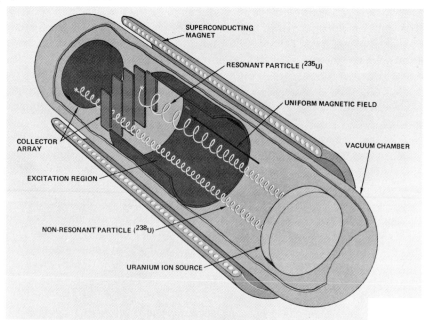

SUPERCONDUCTING MAGNET

RESONANT PARTICLE (^{235}U)

UNIFORM MAGNETIC FIELD

COLLECTOR ARRAY

VACUUM CHAMBER

EXCITATION REGION

NON-RESONANT PARTICLE (^{238}U)

URANIUM ION SOURCE

Source: TRW Corporation, Redondo Beach, California, USA.

Since the rotation of the ions is taking place in the plane perpendicular to the axis of the solenoid, the alternating electric field must also be in this plane so that it can add energy to the ions. Figure 6.26 illustrates this acceleration process. Consider an ion which is moving in the x direction when the electric field is also in this direction. In this situation the field will accelerate the ion and add to its energy, causing it to move in a circle of larger radius but with the same frequency. One half-cycle later, the ion is moving in the opposite direction, but if the alternating field has also changed direction, the acceleration can occur again, and the radius of motion of the ion will be further increased. It is this precise timing of the oscillating frequency of the field to the cyclotron frequency of the ion which is called cyclotron 'resonance'.

The frequency must be stable to an accuracy of 1 per cent or better to distinguish ^{235}U from ^{238}U. This is not difficult to achieve, but it is more difficult to produce a magnetic field which is uniform to better than 1 per cent over a large area. Modern magnet technology can achieve this goal only with relatively sophisticated and expensive solenoid designs (see figure 6.27).

The exciting electric field can be applied in a number of ways, but all must take into account the strong tendency of plasmas to prevent electric fields from penetrating very far into their interior. As was mentioned above, a uranium plasma has a characteristic shielding length of only a few

180

Figure 6.26. Illustration of ion cyclotron resonance
(a)   The uranium ion velocity (v) and electric field (E) are parallel, resulting in an acceleration of the ion and an increase in its radius of curvature.
(b)   The field oscillation is timed to reverse its direction in the same time as the ion reverses its motion; so the ion is again accelerated, and the radius of the orbit increases further. The magnetic field (B) in this figure is directed into the page.

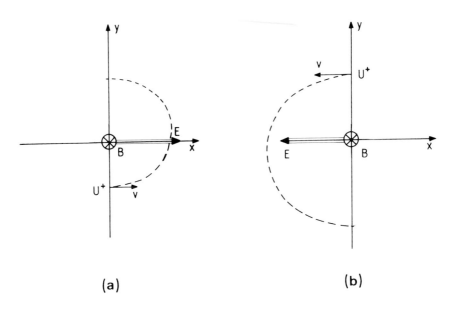

(a)   (b)

micrometres, so externally applied electric fields cannot penetrate deeper into the plasma than this distance. For this reason the electric field must be applied indirectly, either inductively by modulating the axial magnetic field, or by inducing an ion cyclotron resonance wave in the plasma with properly placed electrodes [89b]. Both of these methods create some problems for application, and it is possible that a third, more recently proposed method may prove superior to both [90].

However the electric field is applied, its purpose is to oscillate in phase with the rotating ^{235}U ions. Ions whose cyclotron frequencies are slightly different will fall out of synchronization with the field and will therefore experience no net gain in energy. The effect of increasing the kinetic energy of the ^{235}U ions is to greatly increase their radius of gyration in the field. A typical unexcited ion will have a radius of gyration of the order of a millimetre, but excited ^{235}U ions can be made to move in circles several centimetres in diameter (see figure 6.25). It is not difficult to think of collector designs which would allow most of the ^{238}U ions rotating in their tiny circles to pass through, while capturing most of the ^{235}U ions moving in large circles.

The limitations on density and temperature of the plasma can be inferred from the requirement that the ^{235}U ions be allowed to go through

Figure 6.27. Model of a superconducting solenoid

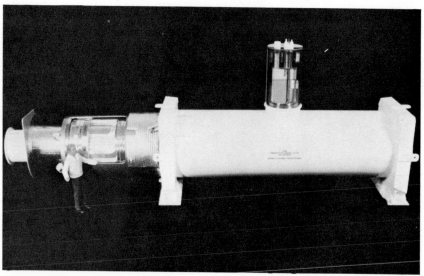

Source: TRW Corporation, Redondo Beach, California, USA.

many cyclotron orbits as they pass from one end of the tube to the other. Because the cyclotron frequency of ^{238}U is so close to that of ^{235}U, the two must make many revolutions in order to ensure that the ^{238}U will be out of phase with the electric field for a substantial number of revolutions. The number of cyclotron periods needed to ensure this is of the order of $M/\Delta M$ or about 80. If the separating tube is about 1 m long and the period of a cyclotron oscillation is 7.7 μs (1/130 kHz) then the axial velocity of the ions cannot be greater than 1 600 m/s. This corresponds to a plasma temperature of about 75 000 K.

The plasma density will be limited by the requirement that a ^{235}U ion must be able to make all of its 80 or so revolutions without suffering a collision with another ion. Such a collision would obviously throw it out of phase with the electric field and cause it to lose energy. This requirement will be satisified if the average ion collision frequency is about 1/80 of the cyclotron frequency, or under 2 kHz. Using a typical collision cross-section of 10^{-14} cm^2 for uranium ions [57c], it can be shown that the ion density in the plasma cannot be greater than 10^{12}/cm^3.

Knowing the ion density in the plasma and the average speed down the tube it is a simple matter to compute the particle flux. This is of the order of 10^{17} ions/cm^2 s. If the plasma beam has a diameter of about 100 cm, if the feed material is assumed to be natural uranium, and if all the ^{235}U ions are assumed to be collected, then the rate of collection of ^{235}U would be 5.7×10^{18} atoms/s or about 70 kg ^{235}U/yr. In order to separate 1 t of ^{235}U per year (enough to make 30 t of reactor fuel) the diameter of the beam would have to be increased by almost a factor of 4, to 3.8 m. This might explain the research and development now being carried out on a superconducting solenoid with a diameter of 4 m [49g].

It must be emphasized that this result is only coincidental, and that all of the numbers in the above calculations are highly uncertain. The eventual values of magnetic field, plasma density and temperature, and collection efficiency will depend on many factors which are either unknown at the present stage of development, or, if they are known, not available in published sources. So the estimates made here should be seen as only rough, order of magnitude approximations. They do, however, fall more or less within the ranges mentioned in the literature [89, 2gg].

There are no reliable data on the enrichment factors achievable with the plasma resonance process. One source suggests "about 10" [33]. A possible lower limit to the acceptable enrichment factor can be inferred from the stated mission of the US Advanced Isotope Separation (AIS) programme, of which the plasma separation process (PSP) is a part. The purpose of the programme is "treatment of uranium tails from conventional [enrichment] plants to produce feed-grade material at costs below natural uranium prices" [91]. This suggests that unless enrichment factors of at least 3.5 per cent in one pass (converting 0.2 per cent feed to 0.71 per cent product) are attainable, the process is unlikely to be considered worth pursuing further.[6]

Very little more can be said about how this process will look if and when it becomes commercially or militarily viable. The size of an enrichment unit is comparable to that for the AVLIS process, since the atomic vapour and plasma densities and flow rates are comparable. Both processes will also have to have similar facilities for materials processing and handling. The plasma process will not require the elaborate laser infrastructure of AVLIS, but in its place will be the need to supply large amounts of liquid helium to maintain the magnets in their superconducting state. The energy consumption of the plasma process will probably be somewhat larger than that of AVLIS, but no firm estimates are available. One source does mention a possible specific energy consumption of "a few hundred kWh/SWU" for anticipated plasma facilities [92].

Plasma centrifuges

There are a number of variations of the plasma centrifuge concept, but all involve the creation of a rapidly rotating plasma. This is achieved by creating an electrical discharge in a direction perpendicular to a strong magnetic field [2hh] (see figure 6.28). The geometry in the figure is simplified for illustration purposes. Actual plasma centrifuges can be much more elaborate [93] (see figure 6.29), and they differ in the arrangements

---

[6] This decision has now apparently been made, at least in the USA (see footnote 4, p. 160).

Figure 6.28. Schematic diagram of one type of plasma centrifuge

A flowing discharge is created by applying a voltage between the anode and cathode. The discharge flows across magnetic field lines created by a solenoid which surrounds the discharge tube. Rotation of the plasma is induced by the mechanism shown in figure 6.23.

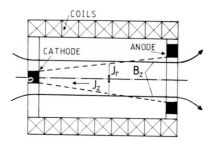

Source: F. Boeschoten and N. Nathrath, "Plasma separating effects", in Villani, S. (ed.), *Uranium Enrichment, Topics in Applied Physics*, Vol. 35 (Springer-Verlag, New York, 1979), p. 294.

of the discharge electrodes, field configurations and methods of feeding in and collecting uranium [15f]. But all share the very great difficulties associated with the complicated behaviour of plasma discharges in magnetic fields. These have been summarized as follows: " . . . so far no gas discharge operated in the presence of a magnetic field is understood well enough that all the properties of the plasma may be calculated. It is even more difficult to obtain reliable information about the role of the neutral particles. Evidently the state of the art in isotope separation with plasmas is directly related to the extent that the used plasma is understood" [2ii].

The isotope separation effect in a rotating plasma depends on the collisions which occur between the two ionic species in the plasma. As was shown in the introduction to this section, ions of different masses have different rotational speeds, so at a given radius they will have a relative velocity with respect to one another. When they collide the heavier species tends to speed up and the lighter to slow down. The result is a tendency for the heavy ions to drift to the periphery of the rotating plasma while the light ions concentrate on the inside [93]. In this way the results are quite similar to the behaviour of UF_6 gas in a mechanical centrifuge, but the physics of the process is far more complicated and difficult to manipulate.

One intuitive prediction which is borne out is that it will require considerably more energy to set the entire plasma into rotation than to selectively excite only ^{235}U ions. Early experiments on isotope separation in krypton gas gave specific energy consumptions 10 times greater than for gaseous diffusion [2jj]. It is suggested that this can be reduced considerably by optimization of design, but the numbers suggested are still substantially larger than those for mechanical centrifuges.

Figure 6.29. Rotating plasma device with curved magnetic field lines

The plasma occupies the shaded area in which an electric field E produces a rotation around the axis of symmetry and a centrifugal force by which ions with different masses are separated.

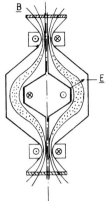

Source: B. Bonnevier, "Experimental evidence of element and isotope separation in a rotating plasma", *Plasma Physics*, Vol. 13, 1971, p. 765.

One suggestion to reduce energy consumption is to use a partially ionized plasma with a low degree of ionization. In this variation the charged particles can be accelerated into rotation, and this rotation can then be transferred to the neutral gas by viscous forces (i.e., collisions) [94]. However, there is no evidence that this idea has progressed much beyond the original suggestion.

Another factor which suggests that the plasma centrifuge may have a difficult time competing with existing methods is the comparison of theoretically obtainable single-stage enrichment factors with those of the mechanical centrifuge. One early estimate for uranium was 1.134 [95a], and one experimental determination gave values in the range of 1.05−1.10 [2kk]. Experiments on krypton gave values in the range 1.05−1.15 depending on the cut [2jj]; and up to 1.10 was achieved in experiments on neon [15g], although this was improved to 1.20 in a later experiment [96]. A Soviet group has achieved an enrichment factor of 1.06 in a $^{129}Xe-^{136}Xe$ mixture [97]. Given that the mass of uranium ions is considerably greater than that of krypton or neon, and that the xenon isotopes had a mass difference of seven units compared to the three of uranium, these latter two results must be upper limits for uranium.

Of course these numbers are not directly comparable with modern centrifuges, because the latter use internal countercurrent flow to enhance the separative effect. This possibility also exists in some forms of plasma centrifuge [95b]. Separation factors can in principle also be increased quite dramatically in plasma centrifuges by increasing the rotational speed of the plasma. There are no bearings to wear out or walls to explode if the rotational speeds are increased to 100 or more times those achievable with mechanical centrifuges. However, such speeds in a plasma require very

185

strong magnetic fields and intense discharge currents, and as the peripheral speed increases, energy consumption and problems with ion-neutral collisions also increase.

Comparisons are also difficult between the plasma centrifuge and cyclotron resonance techniques. While the attainable separation factors for the latter seem to be substantially larger, the former are not limited by collision effects (indeed they require collisions for their operation) so densities and therefore throughputs can in principle be much larger. Densities in plasma centrifuge experiments seem to have been of the order of 10^{15} particles/cm^3 [2 mm], $100-1\,000$ times those of cyclotron resonance.

Summary

There is a kind of historical symmetry about the electromagnetic processes. The first successful enrichment method, the calutron, belonged to this class, and now (35 years later) as the enrichment industry moves towards maturity, these processes are again under serious consideration. None of the modern methods are yet at a stage where either their commercial potential or proliferation implications can be reliably assessed. The numbers quoted and computed in this section (except for the calutron) are certainly the most uncertain of any in this study and must be used with great caution in any comparisons among different types of technology.

VIII. Miscellaneous methods

The previous six sections have described the enrichment processes which have either established themselves as capable of large-scale enrichments or are considered likely enough to do so in the near future that they have been given substantial research and development support. However, this list of a dozen or so methods represents only the tip of the iceberg, the submerged portion of which is made up of dozens, possibly hundreds, more ideas, gadgets, patents, and so on, which have been proposed over the past 30 years.

Wherever a strong commercial or military incentive exists this kind of creativity is certain to follow, and for many years it was believed that the world would beat a path to the door of the nation or corporation which built a better enrichment plant. These expectations have faded considerably in the past few years, but the creativity goes on, and new enrichment ideas appear with surprising regularity.

There is neither need nor space to survey this broad field here, but a few examples might give some idea of the wide variety of proposals which have been made. Since this is intended as a qualitative survey, no numbers will be given or evaluations made. Such numbers or evaluations which do exist can be found in the references.

One group of methods has already been mentioned briefly in section IV. These are the aerodynamic techniques, and there are many variations with names like 'separation probe', 'jet membrane', and 'crossed jet' [26, 28a, 98]. The first directs a supersonic flow of UF_6 in a background gas against the end of an open tube. A shock wave is created at the mouth of the tube, and the separation depends on the differential rate of diffusion of $^{235}UF_6$ and $^{238}UF_6$ across the shock front. The heavier molecules are more likely to penetrate the front, so the tube draws off the depleted fraction, leaving behind an enriched fraction.

The jet membrane is analogous to gaseous diffusion except that the gas is forced to diffuse through another gas rather than through a solid, porous membrane.

The crossed beam technique directs a jet of UF_6 gas at right angles to a jet of some other gas. Collisions occur in the region where the jets intersect, and the lighter $^{235}UF_6$ molecules are deflected slightly more on average than the $^{238}UF_6$. This process is similar in concept to the laser deflection technique described in section VI (p. 172), with molecules substituted for photons.

Another type of separation process relies directly on velocity differences between light and heavy UF_6 molecules. One variant called the 'velocity slip' technique accelerates UF_6 in a light background gas through an expanding nozzle. This is done in pulses, and a rotating velocity selector is used to skim off the lighter, faster molecules as they emerge from the nozzle. Another variation on this theme is the so-called 'garden hose' technique, which again relies on average velocity differences between light and heavy UF_6 and uses a complex spherical sweeping device to skim the enriched fraction [99, 100].

There have been many variations on the centrifuge concept, and a relatively recent one is the 'HAGA radial-separating-nozzle centrifuge', designed and built in Austria [101, 102]. This device is a very complex, multi-chambered centrifuge which relies on vortex and eddy effects at the periphery of a rotating disc to get the separation effect. Very high separative capacities are claimed for rather small devices, but the theoretical and experimental results are very preliminary.

Finally, one of the more recent contributions to the field has been a device that uses a vibrating bar, something like a tuning fork, placed in an atmosphere of UF_6 [103]. When the bar vibrates in one of its resonant modes, the lighter UF_6 tends to concentrate at the locations where the amplitude of vibration is maximum (the antinodes) while $^{238}UF_6$ migrates to the static nodal regions.

Table 6.3. Summary of enrichment process characteristics

Process		Working material	Single-stage separation factor ($q=1+g$)	Stage cut ($\theta = P/F$)	Stage hold-up time (s)	Specific energy consumption (kWh/SWU)	Stage reflux mechanism	Status of technique
Gaseous diffusion		UF_6	1.0040–1.0045	1/2	5–10	2 300–3 000	None	Mature
Centrifuge		UF_6	1.3–1.6	~1/2	10–15	100–300	Internal countercurrent flow	Mature
Aerodynamic	Nozzle	$0.04UF_6 + 0.96H_2$	1.015[a]	1/4	~2	3 000–3 500	Recycle intermediate fraction	Demonstration
	Helikon	0.01–$0.02UF_6 + 0.98 - 0.99H_2$	1.025–1.030[a]	1/20	0.05–0.2	3 000–3 500	None	Demonstration–Production
Chemical	Solvent extraction	Aqueous and organic solutions of U compounds	1.0025–1.0030	1/2	20–30	≤600	Chemical conversions $U(IV)\rightleftharpoons U(VI)$	Pilot plant
	Ion exchange	Aqueous U solution and ion-exchange resin	1.0013	1/2	~1	(400–700)[b]	Chemical conversions $U(IV)\rightleftharpoons U(VI)$	Pilot plant
Laser	Molecular	UF_6-N_2	(5–15)[c,d]	NA	NA	10–50	None	R&D
	Atomic	U vapour	5–15[d]	NA	NA	10–50	Recover and recycle U metal	R&D

188

Electro-magnetic	Calutron	UCl$_4$[e]	20–40[d]	NA	NA	3 000–4 000	Recover and recycle UCl$_4$	Probably un-suitable for large quantities
	Ion cyclotron resonance	U plasma	3.5–10[d]	NA	NA	200–600	Recover and recycle U metal	R&D

[a] In order to use these values in table 5.1, the value of g given here should be multiplied by twice the value of the stage cut.
[b] Estimate based on rough comparison with solvent extraction method.
[c] Estimate based on rough analogy with AVLIS method.
[d] The value entered for the last four techniques is the enrichment factor α (see equations 5.7) rather than the separation factor g.
[e] UCl$_4$ is the usual charge material used in the ion source (see reference 11f and reference 87d). Uranium ions are produced by passing a strong electric discharge through the UCl$_4$ vapour in a heated oven.

This list could go on much longer, but this sample should suffice to support the assertion that the field of isotope separation is far from being exhausted. This suggests that the possibility remains that one day a simple, inexpensive and easily concealable method for uranium enrichment will be discovered. Once the physical principles of a process are understood, efforts to maintain technological secrecy can never be more than temporary stopgaps [104]. And, it can safely be said that where there is a commercial or military will, there is usually a technological way. Any efforts to prevent nuclear proliferation by primarily technological means are therefore constantly threatened with rapid obsolescence.

IX. Summary

The important quantitative data on the techniques described in this chapter are collected in table 6.3. This table is intended for use in conjunction with table 5.1 (p. 113), and the data on separation factor, stage cut and hold-up time necessary for table 5.1 can be taken directly from table 6.3. For an example of how this is done see chapter 5, following table 5.1.

The exceptions to this rule are the techniques with high enrichment factors which are separated from the others in table 6.3 by a double line. Table 5.1 is not applicable to analysis of these, but most of the formulae needed to describe their operation are presented in chapter 5.

References to the sources of the numbers in table 6.3 are not displayed in the table but can be found in the relevant section of this chapter. The few cases where no public data could be found, and where the number represents an educated guess by the authors, are denoted in the table by parentheses. These values should, of course, be used with caution, but this warning should also be made for other numbers as well, especially near the bottom of the table. For those techniques below the double line which are still in their research and development phase, all quantitative estimates must be viewed as highly tentative.

The purpose of the descriptions in this chapter and the organization of the data in table 6.3 has been to provide the reader with sufficient information to make his or her own evaluations of the proliferation implications of each technique. For example, in Part One a possible diversion scheme is mentioned in which a quantity of 3 per cent enriched material is sent to a small clandestine centrifuge facility where it is converted into a small amount of weapon-grade material (90 per cent ^{235}U) with the tails being returned quickly to the large facility from which the diversion took place.

Starting with an assumed requirement of 20 kg of 90 per cent ^{235}U for a nuclear weapon one can first use table 5.1 to compute the quantity of feed material required. In the last column of table 5.1 it is found that 44.5 kg of feed are required for every kilogram of product. So in the present example at least 890 kg of 3 per cent enriched feed would have to be diverted. It can also be seen from the same table that 52.2×20 or 1 044 kg SWU must be performed to get 20 kg of product.

Next assume that the clandestine facility uses centrifuges like those described in table 6.2. If we assume that 4 000 centrifuges are used and that each has a capacity of 15.2 kg SWU/yr, then the facility has a total capacity of 60.8 t SWU/yr and can produce the required 1.04 t SWU in about 1/60 of a year or roughly six days.

Each centrifuge has an inventory of 0.26 g of uranium, so the total inventory of the plant is just over 1 kg, a negligible fraction of the feed required. If each centrifuge has a hold-up time of 13.7 s and a separation factor of 1.51, then table 5.1 can also be used to compute the equilibrium time. Again, referring to the last column and recognizing that a cascade made of centrifuges can be very close to an ideal cascade, we can choose a multiplying factor close to the lower limit of 30, for example, 50. The equilibrium time is then

$$13.7 \text{ s}/(0.51)^2 \times 50 = 2\ 630 \text{ s or } 44 \text{ minutes}$$

Finally note that a conservative estimate of the size of the facility can be made by allowing about one square metre of floor space for each centrifuge. This gives an area of 4 000 m^2, or a square of 63-m sides. A facility of this size would be far easier to conceal than a gaseous-diffusion or chemical-exchange plant.

Part Three

Chapter 7. A history of non-proliferation efforts

I. Early US initiatives

The Baruch–Lilienthal proposal

Among the earliest expressions of concern over the possibility of a future nuclear arms race and a proliferation of nuclear weapons were those of the scientists engaged in the Manhattan atomic bomb project. In the last two war years the eminent Danish physicist Niels Bohr strongly urged negotiations between the United States and the Soviet Union on international control of atomic energy [1]. Also, in 1945, a group of scientists at the University of Chicago Metallurgical Laboratory wrote a report to the US Secretary of War pressing for some form of international control on atomic energy development [2].

Political leaders, in particular Franklin Roosevelt and Winston Churchill, were also aware of the potential significance of atomic energy, not only to the war effort but also to post-war international relations. They understood that the new technology had important implications for both military and industrial development and committed their countries to extensive co-operation in the secret agreements made in 1943 at Quebec. Another part of this agreement was a commitment not to communicate any information to third parties without prior consent. This provision was aimed primarily at preventing as long as possible the acquisition of nuclear 'secrets' by the Soviet Union. After World War II subsequent revelations of spying in the British nuclear programme, and a less co-operative attitude on the part of the Truman Administration and the US Congress, led to severe restrictions on the interchange of scientific and technical information between the two allies. Ultimately, these tensions led to US termination of the atomic co-operation provided for in the Quebec Agreements. However, this did not prevent the USA from later referring to the Agreement when it objected to transfer of nuclear information and possible co-operation between the UK and France on uranium enrichment in the 1950s.

After the actual explosion of two nuclear weapons in Japan at the end of World War II the political leaders of the USA, the UK and Canada explicitly recognized the proliferation risks of an uncontrolled future nuclear development. The source of the troubles, according to their Three-Power Declaration of November 1945, was that "the military exploitation of atomic energy depends, in large part, upon the same methods and processes as would be required for industrial uses" [3a].

In line with these ideas the USA, in 1946, brought to the Atomic Energy Commission of the fledgling United Nations a plan for internationalization of nuclear energy development, the Baruch Plan (see pp. 73–74). However, a crucial provision in the proposal would have allowed the United States to retain its nuclear weapons until full international control of atomic energy had been realized. The implied temporary US monopoly of nuclear weapons was unacceptable to the Soviet Union whose counter-proposal demanded the abolition of all nuclear weapons before establishing international control. This proved unacceptable to the United States.

With the failure of these first attempts at preventing proliferation by internationalization, the USA continued its policy of strict secrecy as prescribed in the Atomic Energy Act of 1946, the so-called McMahon Act. This was a law explicitly designed "to conserve and restrict the use of atomic energy for the national defense, to prohibit its private exploitation, and to preserve the secret and confidential character of information concerning the use and application of atomic energy" [3b].

Accompanying this overt policy of secrecy was a covert attempt by the United States to gain control of all the world's exploitable uranium resources in the belief that this would severely retard other nations' efforts to develop nuclear energy [4]. Although pursued vigorously for several years, this tactic had to be abandoned when it became clear that it was politically unrealistic and that uranium could be found in many more places than were originally known.

A strict secrecy policy

The US secrecy and monopoly policy had no demonstrable effect on the progress of the Soviet nuclear weapons programme. The Soviet Union exploded its first atomic bomb in 1949 using plutonium, and by that time already had an enrichment facility nearing completion. The latter is thought to have started operation sometime around 1950 [5, 6a].

The UK also engaged in both a military and civilian nuclear programme after the war. When the Quebec Agreements were abrogated by the McMahon Act, which excluded the UK along with all other countries from access to restricted information, the British proceeded on their own [6b]. The UK succeeded in constructing and exploding its first atomic bomb in 1952. The fissionable material used was plutonium, produced in British graphite-moderated reactors, which are fuelled with natural rather than enriched uranium.

British requests for US assistance in constructing an enrichment facility had been denied by the United States and the independent establishment of a domestic gaseous diffusion enrichment plant in the UK made relatively slow progress. The plant for the production of highly enriched uranium at Capenhurst was put into operation between 1954 and 1957. By that time British development of a thermonuclear bomb was already under way. For lack of highly enriched uranium, which at that time was the preferred material for use in the trigger of a thermonuclear bomb, the British attempted to develop an H-bomb with a pure plutonium-based trigger. They succeeded in manufacturing such an H-bomb, and tested it successfully in 1957 [6c].

'Atoms for Peace' selective secrecy

The US policy of strict secrecy not only failed to prevent the Soviet Union and the United Kingdom from developing their own atomic bombs, but it also did not prevent other countries, notably in Europe, from developing indigenous nuclear programmes. In contrast to the British programme, which was until the mid-1950s almost completely military, most of the other programmes were aimed exclusively at civilian applications of nuclear energy. However, given the close connection between the technological bases of civilian and military use of nuclear energy, such nuclear activities would inevitably provide more countries with the technical capabilities and materials for manufacturing nuclear weapons.

This 'anarchistic' development and the desire to score a political gain over the USSR caused the USA to change its non-proliferation policy from one of total secrecy and denial to selective secrecy and control by co-operation. In his famous 'Atoms for Peace' speech of 1953, President Eisenhower offered US co-operation to all countries that were or wanted to be engaged in the development of nuclear energy for peaceful purposes. To this end the Atomic Energy Act of 1946 was replaced by the Atomic Energy Act of 1954, allowing the controlled transfer of nuclear equipment, materials, and scientific and technical know-how. The new policy implicitly recognized that possible acquisition of information by the USSR was no longer a problem, since the Soviets had already demonstrated their capability to develop all aspects of nuclear technology, in particular the manufacture of both fission and thermonuclear weapons [7]. Nevertheless, the changes in US policy did not come without a bitter battle within the United States over the need for continued secrecy.

The resulting agreements for co-operation on atomic energy between the USA and other countries contained provisions that the US supplies were to be used only for peaceful purposes. This was to be ensured by means of a US inspection system ('safeguarding').

However, US policy regarding the release of nuclear information and equipment remained restrictive. Sensitive processes, in particular enrichment technology, were considered to be restricted information and kept secret by the United States and not shared with other countries. This policy was motivated both by the fear that a national enrichment plant could provide a country with direct access to weapon-grade uranium, and by the realization that an indigenous enrichment capability would also make the country independent of the United States in satisfying its needs for enriched uranium. The dependence on the USA for enrichment services, as implied by the agreements of co-operation, was supposed to play a central part in implementing the US safeguarding and control of atomic developments in other countries.

Soviet policy towards the Socialist countries resembled US policy in providing these countries with research reactors, nuclear materials and equipment and technical assistance. The Soviet Union also refrained from providing other countries with sensitive technology, such as uranium enrichment, except for the case of the People's Republic of China, which it assisted in building a gaseous diffusion plant. After the termination of Soviet nuclear assistance in 1959 China succeeded in completing the enrichment plant, and in 1964 it became the first nuclear weapon state since the United States to use highly enriched uranium as the fissionable material for its first atomic bombs.

After this traumatic experience, Soviet non-proliferation policy became much stricter. There have been no further exports of enrichment technology by the Soviet Union.

The US policy of combining co-operation and selective secrecy under Atoms for Peace was not completely effective either. France refused to be manoeuvred into a position of dependence on the United States. In particular, it had its own nuclear programme, which at first had mainly civilian objectives but after 1952 became more militarily oriented, notably towards the production of plutonium [8a]. In November 1956, an explicitly military programme was established, including a new protocol for the French Atomic Energy Commission (CEA), charging it with preparing preliminary studies for a nuclear explosion. The CEA was at the same time charged with the responsibility of preparing studies for the creation of an isotope separation plant [6d, 8b]. The production of highly enriched uranium was considered to be of special importance for manufacturing a thermonuclear explosive [6e].

Because of its lack of technical know-how and the very high development and construction costs of an enrichment facility, France had previously asked for British co-operation. At the end of 1954, negotiations had started between France and the UK on the construction of a gaseous diffusion plant, similar to the one in Capenhurst, the first stages of which had just been put into operation. These negotiations were aborted in the beginning of 1955 because of formal objections by the United States based on the Quebec Agreements (see above) [6f, 8c].

After the British refusal of assistance, France, in its desire for an enriched uranium supply independent of the United States, looked for other possible partners. In the negotiations among the six countries of the European Economic Community on the creation of Euratom, starting in 1955, France pressed for consideration of the establishment of a joint European enrichment plant as a major task of Euratom [6f]. This costly undertaking was discouraged by a US offer of cheap and ample enriched uranium supplies to West European countries, an offer made possible by the overcapacity of the large US gaseous diffusion plants, considerably in excess of US military needs.

With no prospect of a joint Euratom enrichment facility, France continued negotiations in 1957 on a trilateral basis with FR Germany and Italy, resulting in an arrangement for co-operation negotiated by the defence ministers of these countries. According to this agreement the FRG and Italy would provide France with financial and technical support for a joint enrichment plant. However, this agreement was rejected in 1958 when General de Gaulle came into power. He did not want to bring FR Germany closer to nuclear armament [6g, 9a].

Finally, in the absence of any support from European partners in establishing a French or joint European enrichment plant, France decided in 1960 to start the construction of a national enrichment facility. This project was strongly opposed by the United States, which had several times since 1958 hinted at the eventual possibility of France obtaining ^{235}U from the USA for French armaments [6h]. Nevertheless, the French gaseous diffusion plant was built at Pierrelatte and put into operation between 1964 and 1967. It produced the highly enriched uranium that was used in the fission trigger of the first French H-bomb, exploded in 1968. The following French thermonuclear test explosion used only plutonium in the fission trigger [6i].

Also during the second half of the 1950s, research on gas centrifuge technology was going on in the United States, the United Kingdom, FR Germany and the Netherlands. In 1960 the US Atomic Energy Commission (AEC) classified this research because of its proliferation-prone character and asked the other countries to act likewise [10a]. Research and development on the centrifuge eventually resulted in the early 1970s in the first enrichment plants outside the present nuclear weapon states, notably in the Netherlands.

In retrospect it is fair to conclude that the selective secrecy of the Atoms for Peace programme did indeed temporarily delay the spread of uranium enrichment facilities to other countries. However, it did not succeed in stopping this process, nor could it prevent France and China from developing their fission and thermonuclear bombs. Two major objectives of the Atoms for Peace programme were to prevent the proliferation of nuclear weapons while at the same time stimulating the application of nuclear energy for peaceful purposes. It has, however, been argued that a fundamental tension exists between these objectives, and that the increased dissemination of nuclear technology and the spread of nuclear reactors greatly enhance the danger of the spread of nuclear weapons [3c].

II. The end of monopoly

Early multinational efforts

The above-mentioned attempts by France to form multinational enrichment consortia were symptomatic of a growing realization by a number of countries that the enormous expense and effort required to develop and construct enrichment facilities were too great for most countries to handle on their own. This realization led to a number of other attempts to encourage co-operation among nations, some of which have been successful.

The motivations for earlier multinational efforts had little to do with non-proliferation objectives. They had much more to do with seeking independence of fuel supply and/or economic advantage. Thus in 1956, after the United Kingdom's refusal to assist France in building a gaseous diffusion plant, six West European countries adopted the establishment of a common isotope separation plant as a major task [11a]. A working group was set up for the purpose of deciding on an enrichment process. This group was later transformed into the Research Association for the Construction of a European Plant and extended to include Denmark, Sweden and Switzerland. At the end of 1957 France took the lead in campaigning for the immediate construction of a gaseous diffusion enrichment plant [8d, 11a], but progress in the joint enterprise was discouraged, mainly as a result of the US offer of low-priced enriched uranium to West European countries [8e]. The secret agreement in 1957 between the Ministers of Defence of France, FR Germany and Italy for a joint effort, including the production of parts of nuclear weapons [9b, 12] (FR Germany and Italy would contribute scientific and financial support for nuclear projects on French territory [9c]), also failed to result in co-operation in an enrichment project. After these failures France decided on the independent construction of the separation plant at Pierrelatte.

Although little progress was made in the 1950s towards multinational collaboration on enrichment, the principle of European multinational co-operation on nuclear matters was institutionalized in the formation of Euratom in 1957. This organization was established by the six West European countries that constituted the European Economic Community (EEC) at the time (Belgium, France, FR Germany, Italy, Luxembourg, and the Netherlands).

Euratom was supposed to co-ordinate, promote and control the development and use of nuclear power in Western Europe and to constitute a framework for obtaining technological support from the United States. The USA supported Euratom as an aspect of its larger interest in European integration and as a potential instrument for implementing US non-proliferation objectives [13a].

On the intergovernmental level the Euratom Treaty of 1957 defines an institutional framework for nuclear developments in the member states. However, it does not prohibit any national activities in the nuclear field. Before the Euratom Treaty was agreed upon much discussion was devoted to the question of whether the Treaty should prohibit the use of nuclear energy in member states for military purposes. At that time France was already engaged in a nuclear programme which would in a few years provide it with a nuclear weapon capability [8f]. Whereas in February 1956 French Premier Mollet stated "We will ask that the member states of Euratom take a solemn engagement to renounce the use of the atom for military purposes" [8g], this position underwent substantial change in the next half year. Under internal pressure from the military and the CEA, among others, Mollet shifted to the position that France's juridical and material capacity to manufacture atomic weapons, if it chose to do so, should not in any way be hindered by French membership in the Euratom community. As a consequence the resulting Euratom Treaty does not forbid member states the use of nuclear energy for military purposes [8h]. The Euratom Treaty does prohibit the diversion of civilian nuclear material for military use, and to this end the Treaty contains provisions for safeguarding nuclear materials and installations used for civil purposes in the member states [14]. Euratom's non-nuclear weapon states are fulfilling their NPT responsibilities by an agreement between Euratom and the IAEA [15]. This agreement was signed in 1973 and entered into force in 1977 after ratification by the member states. It incorporated the essentials of the IAEA's safeguarding provisions of INFCIRC/153 (see chapter 3). Actually, under the agreement Euratom will continue its own inspections, which will be verified by the IAEA [16, 17]. One problem created by this special arrangement is the perception by other nations that it represents what amounts to a self-inspection operation [10b].

The early failures in establishing multinational enrichment facilities did not discourage further attempts in this direction. At the end of 1966 FR Germany and Italy once again brought up the idea of a common European enrichment plant, but neither the UK nor France wanted to allow FR Germany access to their gaseous diffusion techniques. The UK did offer FR Germany a purely commercial share in an eventual enlargement of the Capenhurst plant, but FR Germany was not interested [6j, 11a]. Then, at the end of 1968 the West German, British and Dutch governments announced their decision to co-operate in developing gas centrifuge technology, leading to the first successful multinational enrichment arrangement, Urenco/Centec. This was followed in 1973 by the creation of the second multinational enrichment enterprise, Eurodif.

Urenco/Centec came into being as an *ad hoc* combination of government agencies and public or private industrial firms of the United Kingdom, the Netherlands and FR Germany. It brought together the gas centrifuge isotope separation techniques developed independently in the three countries in an effort to establish a joint enrichment service, independent of the United States. The economic prospects for such an enterprise looked quite favourable in the early 1970s.

Urenco/Centec was established in 1970 under the Treaty of Almelo, a trilateral agreement on the joint development and exploitation of the gas centrifuge process. The corporate structure of the undertaking is rather complex. Centec GmbH is the trilateral society for centrifuge technology, located in FR Germany and established under West German law [18a]. Urenco is the trilateral corporation for management of the enrichment services. A Joint Committee made up of representatives of the three governments deals with 'sensitive political issues', such as safeguards, co-operation with and technology transfer to other countries, location of enrichment plants, and with far-reaching decisions on technological and economic questions. Each of the partners in the Joint Committee has the right of veto [19].

Eurodif was first established by the CEA as a multinational study group to assess the economics of a full-scale gaseous diffusion plant in Europe. The apparent aim was to create an enrichment capacity under French control, primarily to provide the means to satisfy independently its civilian demand for enriched uranium. To sustain this financially risky enterprise, France needed partners both as capital contributors and as assured customers. The economies of scale require commercial gaseous diffusion plants to have large capacities, so both investment and sales must be large to assure profitability.

The original members of the Eurodif study group included private and governmental organizations from Belgium, Italy, Spain, Sweden, FR Germany, the Netherlands and the UK. The latter three countries, after they formed Urenco, withdrew from Eurodif in May 1973 [13b]. In the same year the remaining countries transformed the study group into a private enrichment company, which in 1974 decided to build a large gaseous diffusion plant in France at Tricastin. Sweden subsequently withdrew from the project in 1974 [18b], probably because of the uncertain prospects for future expansion of its nuclear power plant capacity. Sweden's existing enrichment needs were already satisfied by supply contracts with the USA and the USSR. Their 10 per cent share in Eurodif went to Iran in 1975, when the Iranian Atomic Energy Commission and the French company Cogema (a wholly owned subsidiary of the CEA) established the enterprise Sofidif (60 per cent Cogema, 40 per cent Iranian AEO), which acquired a 25 per cent share in Eurodif. Present ownership of Eurodif is divided as follows: France, 50.3 per cent; Italy, 17.5 per cent; Belgium 11.1 per cent; Spain 11.1 per cent; and Iran, 10 per cent (see also p. 215).

Urenco and Eurodif have had mixed success as anti-proliferation measures, but their record has been impressive enough to convince many people of the value of multinational collaborations in the nuclear fuel cycle. Once it was clear that substantial progress was being made in Europe towards independent commercial enrichment enterprises, and once the US monopoly on the supply of enrichment services was broken by a contract for supply of such services by the USSR to France, the USA offered its co-operation in establishing a West European enrichment facility on a multinational base, using US gaseous diffusion techniques. This offer was received with much scepticism in Europe and finally abandoned by the USA as a result of policy conflicts within the US government [11b, 13c, 18a, 20a].

In 1974 the perceived threat of a continuing spread of sensitive techniques to many countries caused the US government (notably Secretary of State Kissinger) to advocate the establishment of multinational or regional fuel centres. The possible transfer of US diffusion and centrifuge techniques was even held out as an incentive [18c, 20b, 21]. The idea of establishing multinational instead of national facilities for sensitive techniques was also endorsed by the London Nuclear Suppliers Club (see below) [22]. However, US enthusiasm cooled as some officials argued that multinational organizations could themselves become vehicles for the further spread of sensitive techniques.

Another endorsement of the multinational concept came in the Final Declaration of the 1975 NPT Review Conference, which took note of the possibility that regional or multinational fuel centres might contribute to non-proliferation goals [23]. The Declaration supported the Regional Nuclear Fuel Cycle Centers Study Project initiated by the International Atomic Energy Agency in 1975 "to determine if multinational fuel cycle centers would have significant advantages for the activities related to the back-end of the nuclear fuel-cycle, in addition to making substantial contributions towards goals of non-proliferation" [24a].[1] The Project report, published in 1977, concluded that implementation of the regional fuel cycle centres was indeed potentially advantageous to serving non-proliferation goals [24b]. Even though this study focused on the back-end of the nuclear fuel cycle, a number of its conclusions are also valid for multinational enrichment arrangements.

Finally, both the US Nuclear Non-Proliferation Act of 1978 [25a] and the International Nuclear Fuel Cycle Evaluation (INFCE) conference (see below) recommended multinational arrangements as important institutional measures for minimizing the proliferation risks of sensitive nuclear techniques. Given this widespread advocacy for multinational fuel cycle collaborations it is important that the existing models (Urenco,

---

[1] The 'back-end' refers to those parts of the fuel cycle such as reprocessing, plutonium storage and radioactive waste disposal which follow the irradiation of nuclear fuel in reactors to produce electricity.

Eurodif and Euratom) be examined carefully to determine how well they carry out their non-proliferation function. This analysis is done in chapter 3.

Commercialization and conflict

By 1970, when the Non-Proliferation Treaty (NPT) took effect, concern about proliferation had decreased considerably, and much more attention was being paid to the apparently great commercial opportunities presented by the expected growth of the nuclear electric power industry. This was certainly the case for uranium enrichment, where, for example, political pressure in the USA increased for turning enrichment activities over to private industry. While advantageous for industry, this step would have had the effect of diminishing the role of enrichment services in US non-proliferation policy. Despite this the US AEC established two programmes to encourage the private sector to develop the capability to build enrichment facilities [26] and started negotiations with interested companies to this end. However, for two important reasons the objective was never realized. First, private industry lost interest in the financially risky enrichment undertaking because of slow-downs in nuclear energy growth; and second, renewed attention around 1975 to the special role of sensitive technology in nuclear proliferation caused the US government to have second thoughts about the benefits and risks of a private enrichment industry [27a]. The Indian nuclear explosion of 1974 played an important part in this reassessment.

Commercial interest in enrichment also grew rapidly in Western Europe in the early 1970s, as shown by the rapid growth of Urenco and Eurodif. Not only did a competitive market in enrichment services, involving several independent enrichment enterprises, arise, but also the transfer of sensitive techniques (notably reprocessing and enrichment methods) began to be included in nuclear package deals between West European countries (in particular, France and FR Germany) and other countries. In addition, these deals were made not only with countries party to the NPT, but also with non-NPT countries. A nuclear package deal in 1975 between FR Germany and Brazil comprised a nearly complete nuclear fuel cycle, including eight nuclear reactors, a fuel fabrication plant and both an enrichment and a reprocessing plant. France contracted to build a reprocessing plant in South Korea (1975) and in Pakistan (1976) [28], and Taiwan also acquired an option to obtain such a facility from France [29a]. This burgeoning trade in sensitive technology, together with the Indian nuclear explosion in 1974, in which material diverted from an unsafeguarded reactor was used to make the explosive, created deep concern, especially in the USA, with regard to the possible consequences about the proliferation problem. It was clear that these sensitive techniques would open the door to direct access to weapon-usable materials. The lead time for manufacturing a nuclear bomb by a country possessing modern enrichment or reprocessing capabilities would in general be very short, once a political decision to obtain a bomb had been taken.

In 1975 these developments resulted in further changes in the US position on several proliferation issues. In particular, the safeguarding of sensitive facilities was no longer considered to be a sufficient barrier against diversion of weapon-usable material and a possible spread of nuclear weapons. According to US reasoning, it followed that the spread of sensitive facilities and technology themselves should be limited. Consequently, the USA put pressure on France and FR Germany not to transfer the above-mentioned enrichment and reprocessing facilities. When it encountered strong resistance in the supplier countries, the USA also put pressure on the receiving countries. As a result South Korea and Taiwan, both heavily dependent on the United States for their national security, cancelled their contracts, but Pakistan and Brazil resisted US pressure. A few years later, in 1978, France suggested to Pakistan a modified reprocessing facility in which plutonium and uranium are extracted together from spent fuel. However, Pakistan showed no interest and finally the French assistance in construction stopped. West German deliveries to Brazil have been retarded by the slow-down in the Brazilian nuclear programme, but construction of the first demonstration enrichment cascade is under way (see figure 7.1).

Figure 7.1. The Brazilian enrichment facility at Resende

Source: E.W. Becker, P. Nogueira Batista and H. Völker, *Nuclear Technology*, Vol. 52, 1981, p. 114.

The London Club (1975)

These developments clearly demonstrated the differences in non-proliferation policy between the USA and a number of West European countries. A situation had arisen in which sensitive technology contracts, the scope of safeguards, and other non-proliferation conditions had become part of the competition for nuclear export contracts. This situation, together with the failure to reach agreement with each of the competitors separately, caused the Nixon Administration to invite a number of supplier countries for talks on these matters. The first closed meetings of this group took place in London in 1975. In the beginning seven countries participated (Canada, France, FR Germany, Japan, the UK, the USA and the USSR), but in 1976 this number was enlarged to 15 and included Belgium, Czechoslovakia, German DR, Italy, the Netherlands, Poland, Sweden and Switzerland.

The London Club meetings were an attempt to arrive at stricter and more uniformly applied non-proliferation conditions on nuclear exports by the various supplier countries. They focused on the special proliferation problems created by the spread of sensitive facilities and technology, implicitly recognizing the insufficiency of the NPT regime for these matters. Apparently the United States and the Soviet Union again recognized a common interest in creating a stricter non-proliferation regime. Neither the United States nor the Soviet Union (since its 'enrichment experience' with China) has transferred any sensitive facilities or technology to other countries. In 1976 US President Gerald Ford announced that the United States would continue its refusal to export reprocessing and enrichment facilities and their technology. The battle lines at the London meetings were drawn between the USA and the USSR on one side, and France, FR Germany and Japan on the other. The United States tried once more within the framework of the London Club to get France and FR Germany to cancel the above-mentioned contracts involving the transfer of sensitive equipment and technology, but the effort was again unsuccessful. The only positive result of this effort was that France in 1976 joined the US embargo on export of reprocessing facilities, but only for future sales [29b]. FR Germany followed in 1977, also exempting its export contract with Brazil. In the meantime the US Congress had passed the International Security Assistance Act, directing the Administration to cut off military and economic aid to countries supplying or receiving reprocessing and enrichment plants and technology [25b, 29c].

In 1976 the London Club agreed on a number of nuclear export guidelines which were made public in 1978 [29d, 30a, 31]. These constituted a voluntary 'gentlemen's agreement' and did not amount to a treaty. The special position of sensitive facilities and technology in the proliferation problem is made clear in these guidelines, the relevant parts of which are summarized and analysed in chapter 3.

III. Recent US initiatives

The anti-plutonium decision (1977)

Although it is not directly related to uranium enrichment, the 1977 decision by the Carter Administration to ban all commercial fuel reprocessing serves as an interesting example of an effort to deliberately avoid a potentially useful process just because of the proliferation dangers associated with it. Its implications are worth examining, because suggestions for similar policies have been made with regard to enrichment technology [32].

In 1977 President Carter followed a recommendation in the Ford-MITRE Report which stated that the USA should defer "indefinitely the commercial reprocessing and recycling of the plutonium produced in U.S. nuclear power programs" [27b]. The Carter Administration decided to restructure its breeder programme "to give greater priority to alternative designs of the breeder other than plutonium and to defer the date when breeder reactors would be put into commercial use", asking other countries to join this policy [33].

The Carter anti-plutonium policy met with strong resistance from other countries, notably from France, FR Germany, Japan and the UK, all of whose nuclear policies were strongly oriented towards the future commercial use of plutonium. The USA was accused of trying to keep a dominant position in the nuclear field, because it was precisely in the breeder programme that the United States was lagging behind the West European countries. It was also alleged that because of its large uranium resources, the USA could tolerate the 'luxury' of a nuclear fuel cycle which did not use plutonium. It was argued that this situation did not hold for other countries. Nevertheless, a few years later FR Germany abandoned its plans for building a large reprocessing plant at Gorleben, officially for internal political reasons, such as resistance from citizen movements. However, there were also strong indications that both the USA and the USSR had urged FR Germany to refrain from building the plant.

France and the UK have continued commercial reprocessing and are even expanding these activities. The USSR is also continuing its breeder programme, probably viewing the spread of reprocessing facilities to be a problem caused by the nuclear export policies of Western countries, something which should not have any repercussions on the Soviet breeder programme. The Soviet Union has not exported reprocessing facilities and requires the spent fuel produced in Socialist countries from Soviet-supplied uranium to be returned to the USSR for reprocessing. However, this requirement is not imposed on West European states who buy Soviet enrichment services.

The Carter Administration's decision to abandon commercial reprocessing in order to avoid the circulation of large amounts of separated plutonium also had a direct impact on its enrichment policy. The anti-plutonium decision included plans to increase the US capacity to produce nuclear fuels, "enriched uranium in particular, to provide adequate and timely supplies of nuclear fuels to countries that need them so that they will not be required or encouraged to reprocess their own materials" [33]. Thus the supply of uranium enrichment services again became an instrument in US non-proliferation policy.

The Nuclear Non-Proliferation Act (1978)

A stricter US non-proliferation policy, which the US Congress had begun to urge under Presidents Nixon and Ford, ultimately won the approval of the Carter Administration. The result in 1978 was the Nuclear Non-Proliferation Act (NNPA) [25c]. The NNPA was in fact the first comprehensive legislative change of nuclear energy policy since the Atomic Energy Act of 1954. The Law gives special attention to the matter of non-proliferation conditions to be included in agreements on nuclear co-operation with other countries and for nuclear exports. These conditions are more or less equivalent to Nuclear Suppliers Group Guidelines. However, in the following provisions the NNPA went even further.

1. Not only were safeguards required on supplied nuclear materials and facilities, but full-scope safeguards were also demanded for non-nuclear weapon states. For exports to nuclear weapon states, safeguards were required on the delivered nuclear items.

2. Prior consent by the USA for retransfer by a recipient country was not only required for 'sensitive' nuclear materials, facilities and technology, but also for all US-supplied nuclear materials, equipment and facilities.

3. Prior consent by the USA was required for reprocessing spent fuel produced from nuclear fuel or with equipment supplied by the USA. In any new agreement for nuclear co-operation the requirement of prior consent by the USA must also be satisfied for further enrichment of US-supplied fuel.

These conditions were supposed to apply not only to future exports, but also to existing agreements. A two-year transition period was provided in the Act to allow renegotiation to bring existing agreements into accordance with the NNPA requirements. If after that period no agreement with the recipient country had been reached on the fulfilment of the export conditions, an export licence could only be issued if specific criteria were met and if failure to approve the export would be "seriously prejudicial to the achievement of United States non-proliferation objectives or otherwise jeopardize the common defense and security" [34a].

According to the NNPA the achievement of US non-proliferation objectives once again rests heavily on assurances of nuclear fuel supply, especially the supply of enrichment services. The Act states that the USA "will provide a reliable supply of nuclear fuel to those nations and groups of nations which adhere to policies designed to prevent proliferation" [34b]. To this end US uranium enrichment capacity was to be increased, a decision previously announced by President Carter in his 1977 anti-plutonium policy. In addition, the USA decided to pursue a vigorous research and development programme on advanced isotope separation (AIS) methods, in order to maintain its leadership in this field. The AIS programme was aimed at developing separation techniques that would make the enrichment of the tails from present enrichment facilities economically attractive, thus extending existing uranium supplies (see p. 183). Finally, the NNPA advocated the establishment of an international fuel authority (INFA) with responsibility for providing fuel services to ensure supply on reasonable terms. These fuel services should be supplied, however, only under strict non-proliferation conditions, such as full-scope safeguards for recipient non-nuclear weapon states. The services should also be available only to countries which do not establish any new enrichment or reprocessing facilities under national control, and which place any such existing facilities under "effective international auspices and inspection" [34c]. The guarantee of an assured fuel supply by such an authority was to help in minimizing the number of enrichment plants under national control, and therefore in limiting physical access to the means of production of weapon-usable material.

IV. Recent international efforts

INFCE (1978–1980)

The Carter Administration's anti-plutonium decision and the NNPA were both unilateral measures, just as the Nuclear Suppliers Club guidelines were the result of a one-sided effort by a group of technologically advanced countries to impose their non-proliferation objectives on other countries. These 'unilateral' actions drew strong protests from other countries and were only partly successful. Therefore the USA also began to look for ways to arrive at a broader international agreement on a non-proliferation regime, stricter than that of the NPT, but at the same time acceptable to more countries. In particular, ways were sought to influence countries engaged in nuclear activities, but not party to the NPT. With this objective in mind President Carter, in announcing his anti-plutonium decision in 1977, called for an International Nuclear Fuel Cycle Evaluation (INFCE) conference. This would investigate the proliferation dangers of various parts of the nuclear fuel cycle and look for more proliferation-resistant alternatives to reduce these risks.

A total of 46 countries and 5 international organizations participated in the INFCE conference, which lasted from 1978 to 1980. Among these countries were several relatively advanced nuclear countries not party to the NPT, such as Argentina, Brazil, France and India. INFCE was organized as a technical conference, in which eight working groups investigated various aspects of the nuclear fuel cycle. The aim was to provide thorough technical and economic analyses to support the development of less proliferation-prone nuclear fuel and reactor strategies for the future.

The most important issues discussed at the INFCE conference centred on two main areas. On one side was the availability of nuclear fuel (resources, prices, international trade) and of various nuclear facilities and technology. On the other side considerable attention was paid to spent fuel management, reprocessing, plutonium management and breeder reactors. Emphasis on these issues reflected the differing interests of the various participating countries. For the USA one of the main interests was to emphasize the proliferation risks of a plutonium economy and to show that commercial reprocessing, plutonium recycling and breeder employment were economically unattractive, at least for the coming decades. Along with this the USA tried to show that technical alternatives were available or could be developed, both for spent fuel management without reprocessing and for a more economical utilization of uranium resources. Other countries stressed the need for an assured supply of nuclear fuel, in both the short and long terms, and the timely availability of the technological means to this end, for example, the breeder reactors. Their attitude in making choices on specific reactor and fuel strategies was to emphasize that "the risk of proliferation must be balanced against any economic, environmental, energy strategy and resource utilization advantages these facilities may have", in which "some risk of proliferation might be considered acceptable" [35a].

In its analysis of the proliferation risks inherent in enrichment technology the Enrichment Working Group established three categories: the diversion of nuclear materials, the spread of technology, possibly leading to the construction of an undeclared or unsafeguarded facility, and the misuse of a declared facility devoted to commercial purposes [35b]. To reduce the proliferation risks of enrichment, INFCE considered the use of the following three methods.

1. International safeguards should be applied to materials and facilities through a system of material accountancy reports, on-site inspection and verification, and various containment and surveillance techniques. Safeguards capabilities were evaluated rather positively at INFCE, and suggestions were made for improvement. It was noted, however, that the only practical experience so far gained is that of Euratom in safeguarding Urenco enrichment plants. This also means that experience has been gained only with gas centrifuge technology. None of the gaseous diffusion plants in France, the UK, the USA, or the USSR have ever been open for inspection (see chapter 3).

2. Institutional measures involving either national or multinational

arrangements were favoured for supervision of plants, technology transfer and nuclear materials. Such measures included classification, export control of equipment and enrichment know-how, and the establishment of facilities under multinational auspices. It was concluded that these institutional measures are partly available and "to some extent have been effective in reducing the risks and concerns which would not be covered by international safeguards alone" [35c]. However, these arrangements were not elaborated in much detail by the Working Group.

3. Certain special features of various enrichment techniques were identified as being potentially helpful in making the clandestine production of highly enriched uranium more difficult. However, opinions differed strongly as to the real influence which these specific technical features might have in a country's decision to construct a small clandestine facility [35d].

The Working Group was aware that these measures could at best reduce the proliferation risks of enrichment activities, but not eliminate them; all enrichment activities remain potentially dangerous. Consequently it stressed that "limitation of the number of plants and development of additional enrichment capacity only in response to needs of a competitive market would be desirable from the perspective of non-proliferation" [35e]. It was concluded that the enrichment market should be competitive, with free access to it by the developing countries, in order that there would be "appropriate flexibility in supply arrangements", reducing for these countries the need to establish their own facilities [35f]. It is also concluded that only a few states in the world are actually in a position to develop commercial-size enrichment capabilities on a national level. Such facilities require a large capital investment, a highly developed technology base and an advanced industrial infrastructure. Of those few states capable of developing national facilities, "those having substantial commercial or industrial incentives to do so would include countries having a large domestic nuclear power program or large indigenous natural uranium resources" [35e]. This, of course, avoids the question of whether other countries might build a small dedicated facility for different reasons.

In its Summary Volume INFCE states its consensus on the relative importance of the above-mentioned three measures against proliferation. The conclusion was that "technical measures have a powerful influence on reducing the risk of theft, but only a limited influence on reducing the risk of proliferation. It is judged that safeguards measures are more important than the technical measures. Potentially more important than technical measures are the institutional measures" [35g]. Such institutional arrangements to reduce proliferation risks would include multinational arrangements for the management of sensitive facilities, an international spent fuel and plutonium storage regime, and international fuel supply arrangements. However, just as in the report of the Enrichment Working Group, these institutional arrangements were not described or analysed in any detail. Such an analysis has been attempted in chapter 3 of this book.

IAEA Committee on Assurance of Supply (CAS)

Partly in response to concerns raised by INFCE, the IAEA has set up three expert consultant groups to study specific aspects of the nuclear fuel cycle. One of these groups is studying the possibility of International Plutonium Storage (IPS), and another International Spent Fuel Management (ISFM). The one most relevant to the enrichment industry is the Committee on Assurances of Supply (CAS) which was created in 1980 to discuss and make recommendations on issues relating to international supply of nuclear material and equipment. The motivation for this committee is the assumption that "assurance of supply and assurance of non-proliferation are complementary" [35h]. The hope is that incentives for establishing national enrichment and reprocessing facilities by a country might be reduced if nuclear fuel supply were guaranteed in accordance with its needs.

V. Concluding note

This brief account of the history of non-proliferation efforts has focused on the role played by uranium enrichment. This industry has been seen both as a cause of proliferation and as a potential means for controlling it, and a wide variety of mechanisms have been attempted or proposed to use enrichment for the latter purpose. In chapter 3 these efforts are categorized and analysed on the basis of the degree to which they involved international collaboration. This variable seems to be a critical one in determining the degree of success of non-proliferation measures.

Chapter 8. The world enrichment picture

I. Introduction

In this chapter the present status of the nuclear and enrichment programmes of all countries presently or prospectively involved in enrichment will be summarized. For this purpose the countries involved have been divided into three categories.

1. Countries or multinational groups with operating enrichment facilities.

2. Countries with relevant R&D programmes, and in some cases more or less definite plans to build an enrichment plant.

3. Countries with possible motivations for developing an indigenous enrichment capability, stemming from either large enrichment needs for their domestic nuclear power programme or their possession of large uranium resources which it might be economically attractive to sell in the enriched form.

The country summaries focus on enrichment status and prospects, and they include other material, such as nuclear energy programmes, interest in plutonium recycling, and breeder reactors, only to the extent that these affect enrichment needs. Each country's attitude towards nuclear proliferation is characterized by combining its status with regard to the Non-Proliferation Treaty with other available data on non-proliferation policies or behaviour.

The results of this survey are summarized in tables 8.2 and 8.3 (see pp. 228–29, 237) where table 8.2 gives the present (January 1981) status of world enrichment, while table 8.3 gives the projections for 1985 and 1990.

In appendices 8A and 8B the enrichment data on individual countries are combined to produce an overall picture of the world-wide status and prospects for enrichment supply and demand.

II. Existing enrichment capabilities

USA

Between 1944 and 1955 the USA built three large enrichment plants, using the gaseous diffusion technique. These plants are located at Oak Ridge, Tennessee (4 730 t SWU/yr); Paducah, Kentucky (7 130 t SWU/yr); and Portsmouth, Ohio (5 190 t SWU/yr) (see figure 8.1). These plants were originally built to produce highly enriched uranium for nuclear explosives. Because of diminished demand for weapon material, they operated at a lower capacity from the mid-1960s until the beginning of the 1970s. Since then the production has again been increased because of increasing demand for nuclear reactor fuel [36a].

Figure 8.1. The gaseous diffusion plant at Portsmouth, Ohio, USA
The US government intends to construct a large centrifuge facility on this same site.

Since the mid-1970s the USA has been enlarging the capacity and improving the efficiency of its three existing plants. This is being done under two programmes, the so-called "Cascade Improvement Program" (CIP; 5 500 t SWU/yr) and the "Cascade Uprating Program" (CUP; 4 600 t SWU/yr) [37]. At Oak Ridge and Paducah the CIP/CUP programmes were scheduled to be completed by the end of 1981. At Portsmouth this will occur in 1983.

In addition to these improvements the United States is also planning a new enrichment plant. This project, which will employ large capacity centrifuges, was announced on 20 April 1977 [38]. The plant will be situated at Portsmouth, and its ultimate capacity is projected to be 8 800 t SWU/yr in 1994. According to the most recent plans the initial capacity will be 2 200 t SWU/yr by 1989, and additional capacity will be built in increments of 1 100 t SWU/yr depending on the demand for enrichment services [39a]. In summary, this means that the US enrichment capacity will increase from 25 300 t SWU/yr in 1980 to 29 500 t SWU/yr in 1990.[1]

Nuclear power generation in the United States is likely to increase from an installed capacity of 54 GW(e) in 1980 to 139 GW(e) in 1990 [40] (see appendix 8A). Corresponding US domestic demand for enrichment services will increase from 6 000 t SWU/yr in 1980 to 15 200 t SWU/yr in 1990 (see appendix 8B). Thus the USA has substantial extra capacity which can be used for export purposes. This extra capacity is estimated at 19 300 t SWU/yr in 1980 and 14 300 t SWU/yr in 1990.

Until the end of the 1970s the USA had a virtual monopoly on uranium enrichment services outside the centrally planned economies. The USA was thus in a position to demand rather strict safeguards in order to ensure that the supplied materials were used only for civilian purposes. As more suppliers enter the market, the effectiveness of this leverage is becoming more limited. Customers who do not want to accept the US conditions can now negotiate with other suppliers. Their chances for buying enrichment services from other suppliers will be good, because at least till the beginning of the 1990s there will be a substantial overcapacity of enrichment services (see appendix 8B).

However, the United States still has two important advantages in its attempt to maintain world leadership in uranium enrichment. First, the USA has large domestic uranium resources, some 30 per cent of the (non-CPE) world's demonstrated and estimated totals. In 1979 the US production of uranium was almost 40 per cent of the world's total [41]. This means that in many cases the USA can link together the sales of natural uranium and of enrichment services. The second advantage is the US price for enrichment services, which is substantially lower (some 20−30 per cent) than those of her European competitors [42a]. This is possible because the existing plants have already been written off under military programmes, and because the US Department of Energy (DoE) is required by law to recover only uranium enrichment costs. There are no profits, no taxes, no insurance or similar business costs in the DoE charge [42a]. With respect to the planned gas centrifuge plant the Department of Energy expects that costs per SWU might be 40−60 per cent lower than those of its European competitors [42b]. Because of these advantages the USA still sees its enrichment services as an important instrument of its non-proliferation policy. A commitment has been made to substantial increases in capacity, and a serious search is under way for new, more

---

[1] As this book goes to press the proposed centrifuge facility at Portsmouth has come under attack in the US Congress as the result of an unfavourable report by the US General Accounting Office (GAO). Citing the existing oversupply of enrichment capacity and the potential of even more efficient future methods (lasers), the GAO has strongly questioned the need for a centrifuge plant. Since the project will eventually cost $7−billion, it has become a prime target for budget cutters in the Congress [130].

efficient enrichment techniques under the Advanced Isotope Separation (AIS) Programme [43] (see footnote 4, p. 160).

USSR

The USSR has a gaseous diffusion plant with an estimated capacity of 7 000–10 000 t SWU/yr located somewhere in Siberia [44a]. Nuclear generation of electricity has a high priority in the Soviet Union, which by the end of 1980 had 26 operating reactors with a total capacity of 11.5 GW(e). Ten more reactors with a total capacity of 8.9 GW(e) were scheduled to begin operation before the end of 1981, although by that time no information was available on the actual status of these reactors. About 65 per cent of Soviet capacity is in the form of gas-cooled reactors which do not require enriched uranium [40]. Soviet energy planners project additions of more than 80 GW(e) of nuclear capacity by 1990. These plans are not subject to many of the uncertainties present in Western economies, since there is virtually no public discussion of the risks and benefits of nuclear energy in the Soviet Union. Moreover, the centrally planned economy removes many of the economic constraints connected with the need to make revenues equal or exceed costs.

The USSR has domestic uranium resources (production estimated at 7 000 t/yr), and uranium is also imported from Eastern Europe, especially Czechoslovakia (estimated imports equal 10 500 t/yr). Since 1946 the USSR is reported to have stockpiled approximately 200 000 t of uranium [45a]. Despite this large reserve, the Soviet government cites a potential shortage of uranium as a justification for its intensive research and development effort on fast-breeder reactors. According to current plans commercial breeders should be operating after 1990, and the breeder will become the dominant reactor type in the next century [46a].

Since the break in nuclear co-operation with China in 1959, the USSR has held to a nuclear export policy with tight controls against weapon proliferation. The USSR has limited its nuclear exports to reactor types considered to be proliferation-resistant and has forced client countries to accede to international non-proliferation protocols [45b]. Only countries within the Socialist world and Finland have received nuclear aid so far. Uranium is enriched in the USSR and the spent fuel elements from Soviet-supplied reactors in other countries must be returned to the USSR for possible reprocessing and storage [45c]. However, this condition does not apply to enrichment services supplied to West European countries.

Recently the USSR embarked on a new phase in its nuclear export policies intended to increase its share of the nuclear export market. In particular, the Soviet Union is aiming for commercialization of its 440 MW(e) light water reactors, and recent sales have been made to Cuba and, possibly, Libya [45d]. The USSR is also very active in the West European enrichment market. The 1980 domestic and East European

214

demands for enrichment are estimated to be between 1 000 and 1 700 t SWU/yr (see appendix 8B). This has allowed the Soviet Union to devote some of its surplus enrichment capacity to profitable sales of toll enrichment services to Western Europe. Contracts for these services amount to 37 800 t SWU for the 10-year period 1980–1990 [35i].

The USSR advocates the establishment of international or regional fuel cycle centres under IAEA control, as well as other measures to internationalize the processing and storage of nuclear materials, including plutonium [47a].

Eurodif countries

Eurodif is a multinational uranium enrichment company in which five countries participate [48a] (see table 8.1). Eurodif is constructing the gaseous diffusion plant at Tricastin with a projected capacity of 10 800 t SWU/yr. At the end of 1980 a capacity of 6 000 t SWU/yr was already operating, and in 1981 this was probably extended to 8 400 t SWU/yr. Full capacity was scheduled to be reached in mid-1982 [50].

The Eurodif countries have also created a second multinational enrichment company, called Coredif. The shareholders in Coredif are the same as in Eurodif, although the division of the shares is somewhat different, resulting in greater participation by France and Iran. In the original plans Coredif was to build a plant of the same capacity as the Eurodif plant, of which the first 5 000 t SWU/yr were expected to be available in 1985 [44b]. However, the revolution in Iran, which has at least temporarily stopped the Iranian nuclear programme, as well as substantial cutbacks in the Italian and Spanish nuclear power programmes, suggest that construction of such a new plant is unlikely before the late 1980s at the earliest [51].

Table 8.1. Current ownership of shares in Eurodif

France	35.28%[a]
Sofidif (60% France + 40% Iran)[b]	25.00%
Italy	17.50%[a]
Belgium	11.11%
Spain	11.11%
Total	**100.00%**

[a] The figures for France and Italy from reference [48] have been adjusted to account for the later sale of one-quarter of Italy's share to France (see the following discussion and reference [49]).
[b] 'Sofidif' stands for Société France–Iranienne pour l'Enrichissement de l'Uranium par Diffusion Gazeuse.

Owing to its lack of domestic coal and oil resources France has committed itself firmly to nuclear power. Most of the electricity-generating plants under construction or planned for the near future are nuclear reactors. By the end of 1980 France had almost 13 GW(e) of nuclear power in operation. In 1990 this capacity is scheduled to be over 52 GW(e) [40]. Most of the French reactors are light-water moderated (LWRs) and need low-enriched uranium as fuel. The French demand for enrichment services could therefore increase from 1 700 t SWU/yr in 1980 to 5 450 t SWU/yr in 1990 (see appendix 8B). In 1981 the newly elected French President Mitterand announced that he would cut back the French nuclear programme so that only the reactors that were already under construction would be completed [125]. In comparison with the plans under his predecessor Giscard d'Estaing, this could mean a reduction in planned capacity of almost 12 GW(e) by the late 1980s [40]. On 30 July 1981 the French Council of Ministers decided to suspend the construction of five reactors [126]. In the past the French enrichment demand was supplied by the USA and the USSR. However, the French participation in Eurodif means that in this respect France will become independent of foreign suppliers because the French share will equal a capacity of 5 430 t SWU/yr when the Eurodif plant reaches full capacity in the early 1980s.

Besides its present and future LWRs, France is also putting great effort into developing fast-breeder reactors (FBRs). It has operated a 233 MW(e) FBR, called the Phénix, since 1973, and a 1 200 MW(e) FBR, the Super Phénix, is now under construction in co-operation with Italy and FR Germany [44b]. This is scheduled to be completed by the end of 1983 [40]. However, President Mitterand has announced that after completion of the reactors presently under construction, France will refrain from further use of the breeder reactor [125]. France also has commercial reprocessing facilities at Marcoule and La Hague and plans to increase these capacities in the 1980s [44b]. French uranium resources are estimated at 55 000 t of uranium. Thus, France has developed a complete domestic nuclear fuel cycle.

In addition to the Eurodif facility at Tricastin, France still operates a gaseous diffusion plant (capacity between 300 and 600 t SWU/yr) at Pierrelatte [44c, 52]. As noted in the previous chapter this plant was built for military purposes. It also provided France with the opportunity to develop its gaseous diffusion technology before using it for the Eurodif plant. France is also carrying out substantial R&D on other enrichment techniques, including gas centrifuges, aerodynamic methods, chemical exchange and lasers [53a]. With respect to chemical exchange it is considering the construction of a pilot plant with a capacity of 50−100 t SWU/yr, which could start operation before 1985 [54a]. France is a major exporter of nuclear technology and hopes to increase its share of the world market. Since it is in a position to assure fuel supplies from its own uranium resources and access to resources in its former colonies, Gabon and Niger [53a], France is in a more favourable position than other potentially major nuclear exporters, such as FR Germany and Japan. France is not a party to the NPT but it has stated that it would act as if it were [55a]. It participated under US pressure in the 'London Nuclear Suppliers Group' (see chapter 7) which agreed on guidelines for the export of nuclear materials and technology. France is often criticized for its nuclear exports, especially its reprocessing technology. Deals with South Korea, Taiwan and Pakistan have been cancelled under pressure, mainly from the USA [29a, 56a].

Italy

Italy is a highly industrialized country that depends heavily on foreign energy supplies. In the beginning of the 1970s, it had plans for a large nuclear power programme to reduce this dependence, but owing to economic and reactor-siting problems these plans have been sharply curtailed [44d]. By the end of 1980 Italy had about 1.4 GW(e) of nuclear capacity in operation and about 4 GW(e) planned or under construction, most of the reactors being light-water moderated [40]. Its original 25 per cent share in Eurodif would have been sufficient to make Italy independent of foreign sources of enrichment services, which until now have been supplied by the United States and the Soviet Union. The current size of Italy's nuclear power programme requires only a few hundred tonnes SWU per year. Its share in Eurodif, however, represents a commitment to several thousand tonnes SWU per year [57]. Therefore, it is not surprising that Italy has looked for ways to reduce its commitment to Eurodif. One-quarter of the original Italian commitment has already been sold to France, and negotiations are in progress for the sale, again to France, of the remaining 17.5 per cent Italian share in Eurodif [49]. In addition, Italy has bought from the United States 20 000 t depleted uranium from US enrichment plants. These contain 0.3 per cent ^{235}U, and Italy will use its share of the Eurodif capacity to re-enrich this uranium to the concentration of natural uranium [58]. If a tails assay of 0.2 per cent is assumed for the Eurodif plant, then the 20 000 t depleted uranium will allow Italy to produce about 3 900 t 'natural' uranium. This will require about 3 200 t separative work.

It has been suggested that this uranium might be further enriched in the USSR [59]. For the years 1979–1983 Italy has a contract with the USSR for 4 225 t SWU [57a], which would require about 5 400 t natural uranium, assuming a concentration of 3 per cent ^{235}U in the enriched product and a 0.20 per cent tails assay. About three-quarters of this natural uranium requirement could be produced from the depleted uranium bought in the USA. While producing natural uranium by means of upgrading depleted uranium may be expensive when compared to directly buying natural uranium, Italy may find it economically more attractive than to use its Eurodif capacity for producing low-enriched uranium, which given the global overcapacity of enrichment services may be difficult to sell.

Italy has pursued R&D programmes on uranium enrichment for several years, especially on gaseous diffusion and gas centrifuges. Experimental work has also been done on basic aspects of uranium enrichment by lasers [53b]. Since the decision to participate in Eurodif, Italian work on gaseous diffusion has been primarily directed towards use in Eurodif facilities [48b].

In addition to its enrichment activities, Italy is also carrying out research on fast-breeder reactors by means of its one-third participation in the Super Phoenix Project (see the section on France above). Italy also has two small reprocessing pilot plants [60]. There has been some discussion of constructing a 600 t/yr plant, but no serious plans have as yet materialized [44e].

Italy has ratified the NPT and participated in the London Nuclear Suppliers Club (see chapter 3). Recently, however, international concern has arisen about the sale to Iraq of 'hot cells', which can be used to retrieve plutonium from irradiated fuel elements [61].

Spain

In the mid-1970s Spain imported about 70 per cent of its energy. Rapid growth of electric power consumption and rapidly rising oil prices caused the Spanish government to adopt a heavily nuclear National Power Supply Plan in January 1975 [44f]. By the end of 1980 Spain operated a nuclear power capacity of 1.07 GW(e) and a capacity of 13 GW(e) was planned or under construction, with most of the reactors using low-enriched uranium for fuel [40]. This means that Spain has a considerable need for enrichment services which are contracted to be supplied by Eurodif, the USA and the USSR [57a]. Spain's 11 per cent participation in Eurodif means that it can become less dependent on foreign suppliers, although this share is not sufficient for all of Spain's projected enrichment needs in the late 1980s. These would require about 13 per cent of the Eurodif capacity. Problems with the USA over US conditions for supplying nuclear fuel caused considerable dissatisfaction in Spain [44g], and Spanish nuclear policy seems to be aimed at gaining independence from the United States.

Spain's uranium resources are estimated at some 10 000 t uranium [41], and current production is a few hundred tonnes of uranium per year. Spain has a contract for reprocessing spent fuel from its Vandellas power reactor in the French reprocessing facility at Marcoule [62]. No plans for indigenous commercial reprocessing plants or breeder reactors are known.

Spain has not signed the NPT, citing its security needs and hesitation about adverse economic effects on its well-developed nuclear programme

[55b]. Spain claims, however, that it has always scrupulously adhered to its safeguards agreements with the USA and the IAEA [44g]. Recently it was reported that the IAEA safeguards system would be effectively extended over all of Spain's nuclear fuels and installations. An exception is made for the reprocessing of Spanish fuel at the French Marcoule plant, because France refuses to allow the IAEA to inspect this plant [62].

Belgium

To reduce its dependence on imported primary energy, Belgium has embarked on a relatively large nuclear power programme. By the end of 1980 LWRs with a power output of 1.65 GW(e) were operating, and a capacity of 3.8 GW(e) was under construction and scheduled to be completed by 1984 [40]. This means that from the mid-1980s the Belgian demand for enrichment services will be around 600 t SWU/yr, to be supplied under contracts with Eurodif, the USA and the USSR [57]. Belgium's 11 per cent share in Eurodif is sufficient for its enrichment needs throughout the 1980s. Belgium is also conducting a small research programme on laser isotope separation [63−65].

Belgium has a research programme on breeder reactors and has a 15 per cent share in a 300 MW(e) fast-breeder reactor being built on West German soil in co-operation with FR Germany and the Netherlands [44h]. However, in mid-1981 it was reported that completion of this reactor was endangered by financial troubles [127]. Belgium also has a small reprocessing plant, which was originally operated by several West European countries in the context of the OECD/NEA, but which has been closed down since 1974. However, the Belgian government has agreed to take over the plant and plans to reopen it [66]. Belgium has ratified the NPT.

Iran

Under the Shah, Iran had planned an ambitious nuclear energy programme of 25 GW(e), capable of generating half of Iran's 1977 electricity demand, by 1997 [67a]. Most of the planned reactors were to be light-water moderated. Thus Iran anticipated a substantial need for enrichment services starting in the 1980s. Through its participation in Eurodif and Coredif Iran could satisfy its enrichment needs in an independent way. However, since the revolution, construction work on nuclear power plants in Iran has completely stopped. In the beginning of 1979 two reactors ordered from France were cancelled, and in mid-1979 the West German reactor constructor KWU stopped construction at another plant, which was about 80 per cent completed [68, 69]. In 1980 it was reported that the current Iranian regime had told its European suppliers that it did not recognize the contracts of the previous government for the purchase of reactors [70]. This also led to a discussion in Iran on whether it should sell its Eurodif share. In early 1980 the Iranian Foreign Minister announced that Iran wanted to sell [71], but he was contradicted shortly afterwards by the President of the Iranian Atomic Energy Commission [72]. At the time of writing, Iran seems willing to sell its share in Eurodif but is taking no initiative towards doing so. However, financial problems between Iran and Eurodif may put extra pressure on Iran to sell out [73]. Iran has no known uranium resources. It has stockpiled uranium ore (how much is not reported) and is still exploring for domestic deposits. Thus it seems that Iran has not completely given up its nuclear power activities.

Iran has signed and ratified the NPT; but the war between Iran and Iraq may affect Iranian attitudes towards the Treaty. The suspicion that Iraq is seeking a nuclear weapon capability, possibly in co-operation with other Arab states, could force Iran to reconsider its non-proliferation policies. In October 1980 it was reported that Iranian warplanes had bombed the Iraqi nuclear research centre near Baghdad, but a French-supplied 70 MW(e) research reactor under construction was reported not to be damaged [74a]. Other reports have suggested that this attack was carried out by Israeli planes carrying Iranian markings [62], and the subsequent Israeli attack lends some credence to this theory.

Urenco

On 4 March 1970, the Federal Republic of Germany, the Netherlands and the United Kingdom signed the Almelo Treaty, creating the Uranium Enrichment Company, Urenco. The shareholders in Urenco are: British Nuclear Fuels Ltd (BNFL), a 100 per cent state-owned corporation; Ultra-Centrifuge Nederland (UCN), 98 per cent state-owned; and Uranit, a 100 per cent privately owned West German firm. The parties to the Almelo Treaty agreed to co-operate in the development of the gas centrifuge process for uranium enrichment and on the construction and operation of uranium enrichment facilities, based upon the gas centrifuge, for nuclear energy application [19]. In 1980 two demonstration plants at Capenhurst (UK) and Almelo (the Netherlands), each with a capacity of 200 t SWU/yr, were completed. Additional plants, which will bring Urenco's total capacity to 2 000 t SWU/yr, are now under construction and are scheduled to reach full production in the late 1980s [39b]. These construction activities are now underway at Capenhurst and Almelo, but there are also serious plans to build a plant at Gronau (FRG). When these plans are realized all three Urenco partners will have an enrichment facility on their own soil.

FR Germany

FR Germany has a substantial nuclear power programme. By the end of 1980 it had 8.6 GW(e) in operation and almost 13 GW(e) was planned or under construction [40]. Most of these reactors are fuelled with slightly enriched uranium, for which FR Germany is largely dependent on foreign sources. It has hardly any indigenous uranium resources, and until recently it depended on the USA and the USSR for its enrichment needs. The West German enrichment demand will increase from 1 000 t SWU/yr in 1980 to 2 250 t SWU/yr by 1990 (see appendix 8B). The FRG is making a substantial effort to develop a more independent nuclear fuel cycle. In order to make more efficient use of nuclear fuel it has substantial R&D programmes on reprocessing and breeder reactors. In co-operation with the Netherlands and Belgium it is constructing a 300 MW(e) fast-breeder reactor on West German soil near Kalkar. By the end of 1980 this project was scheduled for completion in 1986. However, financial difficulties were reported in 1981 [127]. To help supply its present reactors the FRG is developing its own enrichment capacity within the Urenco troika.

Because of the political sensitivity of the FRG acquiring an independent enrichment capability it was decided in 1970 that Uranit would build the West German part of the Urenco capacity in Almelo, some 25 km west of the Netherlands–West German border. However, by the end of the 1970s FR Germany had made it clear that it intended to build an enrichment plant on its own soil [75]. The location finally chosen was at Gronau, some 30 km from the Almelo plant, and the planned capacity is 1 000 t SWU/yr [128a]. A first section with a capacity of 400 t SWU/yr is planned to come into operation in the mid-1980s [128b]. Once the facility is built, FR Germany will be very close to an independent enrichment capability, since a major provision of the Almelo Treaty allows any of the members to withdraw from the consortium with one year's notice after 1980.

The FRG has an extensive research and development effort in uranium enrichment in addition to the centrifuge; other major efforts are in the jet nozzle and laser processes [36b]. Research on jet nozzle technology does not appear to be motivated by domestic enrichment needs. Nevertheless, this technology plays an important role in West German nuclear policy. With no indigenous uranium resources the Germans may be tempted to offer jet nozzle technology in negotiations with other countries in exchange for uranium resources. This is precisely what happened in the major agreement between FR Germany and Brazil in which the former sold Brazil an entire nuclear fuel cycle and received an option on portions of Brazilian uranium resources [76a]. A second example may be the co-operation between FR Germany and South Africa on uranium enrichment [77, 78]. It has already been noted (chapter 3, section I) that at least in the early stages the development of the South African Helikon process was supported by West German technical assistance.

Moreover, South Africa has substantial uranium resources. More than 30 per cent of West German uranium imports will be supplied by South Africa in the period from 1981 to 1990 [79a].

UK

By the end of 1980 the UK had 33 nuclear power plants in operation with a total capacity of about 8 GW(e), and a further capacity of about 6.3 GW(e) was under construction [40]. Most of the operating reactors consist of the so-called gas-cooled type, which use natural uranium as a fuel. However, the majority of the reactors under construction are of the so-called 'advanced' gas-cooled type. These reactors use slightly enriched uranium as fuel. Thus for its future reactors the UK will need substantial amounts of enrichment services.

The UK has a complete but not entirely self-sufficient nuclear fuel cycle, including a reprocessing plant, a research fast-breeder reactor and a 400 t SWU/yr gaseous diffusion plant for uranium enrichment [44i]. This capacity is, however, not enough to satisfy projected British needs in the 1980s, especially in view of plans to close down the existing facility, possibly in 1985 [48c]. As a substitute the UK will expand its enrichment capability within the Urenco framework [80]. Research on laser enrichment techniques is also being pursued [54b].

Netherlands

In 1970, when Urenco was founded, the Netherlands had plans for a large nuclear power programme, some 35 GW(e) by the year 2000 [81]. Since then these plans have been drastically reduced, and in 1980 the total Dutch generating capacity was only 500 MW(e) in two facilities [40]. Even these two plants are in danger of being shut down, given the very strong anti-nuclear energy movement in the Netherlands. In these circumstances there is no great need for enrichment services in the Netherlands, and those which do exist are already being supplied under a long-term contract with the United States [57a]. Therefore, at least for the next decade, the Netherlands will use its Almelo centrifuge facility only for export purposes, almost entirely to its Urenco partners. With respect to breeder reactors the Netherlands has a 15 per cent share in a reactor which is, in co-operation with FR Germany and Belgium, under construction on West German soil. However, completion of this reactor has become uncertain due to financial problems [127].

All three Urenco partners have signed and ratified the NPT. All have frequently expressed their opinion that measures should be taken in order to reduce the risk of proliferation of nuclear weapons. The three Urenco partners are also members of the NATO alliance, and the UK is a nuclear weapon state. FR Germany and the Netherlands have US nuclear weapons located on their territories. Although these 'tactical' nuclear weapons can only be released by the US President for use in a conflict in Europe, the Netherlands and West German armies participate in a collective NATO defence strategy and thus will have access to nuclear weapons in such a contingency.

With regard to uranium enrichment, the Treaty of Almelo includes a Joint Committee, consisting of representatives of the three governments, which has to approve unanimously all Urenco export contracts [19]. However, some policy differences among the partners can be observed. An interesting example is provided by the West German contract with Brazil in 1975. FR Germany intended to sell gas centrifuge technology to Brazil, but the Netherlands made clear that they would veto the proposal on the basis of their own non-proliferation policies. Another example concerns the Brazilian wish, announced in 1976, to buy uranium enrichment services from Urenco. The order covered the supply of 2 000 t SWU in the 1980s, some 10 per cent of Urenco's contracted deliveries at the time. The Brazilian contract would be the first Urenco contract with a commercial price for enrichment services. Previous contracts with West German and British utilities involved prices set some 25 per cent lower. The UK and FR Germany had few hesitations on the Brazilian contract, but the Dutch expressed strong reservations. The guarantees that Brazil would not use either the supplied enrichment services or the plutonium that would be bred in its power reactors for nuclear explosives were considered to be insufficient by the Netherlands. Further reservations stemmed from the fact that Brazil was not a party to the NPT. The Netherlands government favoured an International Plutonium Storage (IPS) arrangement controlled by the IAEA for the irradiated Brazilian fuel elements. The Netherlands demanded that at the time of the first delivery to Brazil sufficient indications should exist that such an arrangement would be created. Dutch hesitation on this contract created considerable resentment among its Urenco partners.

This dispute had important effects on Urenco co-operation. In the early stages of the discussion of the Brazilian contract FR Germany and the UK stated their intention to guarantee a supply of enrichment services to Brazil, and if necessary to nullify Dutch objections by terminating the Urenco co-operation in 1981 (the earliest date possible under the Almelo Treaty). FR Germany also threatened to withdraw from the Treaty unless the right of veto in the Joint Committee was rescinded. In 1978 the Netherlands Parliament agreed to sell to Brazil, but with the reservation that by the time of the first shipment of enriched uranium from the Almelo plant to Brazil there should be sufficient guarantees against possible misuse of the supplied materials for military purposes. If, according to the Netherlands Parliament, this should not be the case, then an export licence for the enriched uranium would be refused. This provision was later circumvented when in 1981 Urenco announced that the shipments to Brazil would take place from the Urenco plant in Capenhurst, UK, so no Dutch export licence would be needed.

China

China has no nuclear power plants but is planning a large programme. By the year 2000 a capacity of 15 GW(e) is projected, with 2−4 GW(e) ready by 1990 [82]. China intends to develop a complete domestic nuclear fuel cycle. Plans have been announced for the construction of three prototype power reactors, including two 100 MW(e) heavy-water reactors and a 300 MW(e) LWR [74b]. In order to accelerate the development of its domestic technology, China is interested in purchasing a few reactors from abroad. China would prefer to purchase US technology [83], but a major obstacle is its lack of hard foreign currency [67b]. China is also interested in other suppliers, and at the end of 1980 it was announced that China and France had reached an 'accord in principle' on the French supply of two 900 MW(e) LWRs [84]. There have also been negotiations between the British colony of Hong Kong and the Chinese province of Guangdong on the joint construction of one or two nuclear power plants on Chinese soil [85].

China operates a small gaseous diffusion plant with a reported capacity of 180 t SWU/yr at Lanchou, completed in 1963 [86]. Much of the know-how for this facility was provided by the Soviet Union during the period of Sino–Soviet nuclear co-operation which ended in 1959 [87]. China also conducts research on gas centrifuges and laser enrichment [88], and has a small heavy water production plant of 40 t/yr capacity [89]. In the long term China plans to develop both breeder reactors and fusion power. In this connection an agreement was signed with Italy in 1980 which allows Chinese technicians to work in Italian laboratories on fusion research, breeder reactor construction, fuel fabrication and radio chemistry [82]. China already has some reprocessing capacity (information on its size is not available), and plans to build a plant with a capacity of 80−100 t/yr [89].

With regard to uranium resources a Chinese official has stated that "we have enough resources [to supply the planned 15 GW(e) capacity] for a long time" but that any demands beyond that capacity would require more resource development [82]. If "a long time" is assumed to mean the lifetime of a reactor (approximately 25 years) this would indicate resources of at least 50 000 t of uranium.

China is a nuclear weapon state. It has refused to sign the NPT because of a stated belief that for some countries it may be necessary to acquire nuclear weapons both for national security and in order to withstand economic and military pressures from the USA and the USSR [55c]. Moreover, China takes the position that the NPT discriminates against non-nuclear weapon states and puts restrictions on the peaceful uses of nuclear energy. This view suggests that China will not accept safeguards on its own nuclear facilities.

Japan

Japan is a highly industrialized country with hardly any indigenous energy resources. It considers nuclear power to be the most reliable future energy option. In 1980 Japan had 22 nuclear power stations with a total capacity of about 14.5 GW(e) in operation. Twelve more plants with a capacity of almost 10 GW(e) are planned or under construction [40]. By far the majority of current and planned Japanese reactors are light-water moderated, and the introduction of the fast-breeder reactor is not expected before the beginning of the 21st century. Therefore, Japanese requirements for uranium enrichment services are expected to increase from 1 600 t SWU/yr in 1980 to 2 650 t SWU/yr by 1990.

Currently the Japanese enrichment demand is supplied through long-term contracts with the United States amounting to almost 70 000 t SWU in the period from 1979 to 2000. Another contract has been signed with Eurodif for the supply of 10 000 t SWU between 1980 and 1989 [57a]. These contracts provide a supply sufficient to operate 35 GW(e) of capacity for about 20 years. Therefore, even though the growth of nuclear capacity in Japan is expected to be substantial, there is a sufficient supply of uranium enrichment services, at least for the coming decade, from existing foreign contracts. The Japanese government foresees a shortage in uranium enrichment services somewhere in the 1990s. An important consideration for Japan with respect to new contracts with foreign suppliers is that the uranium market is not one of perfect free trade. An oligopolistic supply structure of enrichment services coupled with possible political complications resulting from non-proliferation policies of governments could affect pricing and supply conditions. Mainly because of these reasons Japan does not want to depend entirely on foreign supply sources and is developing an enrichment capacity of its own [39c]. The Japanese history of R&D on uranium enrichment is rather long, beginning with a little known effort during World War II [90]. The first successful centrifuge for uranium enrichment was constructed in 1959 [39d]. In 1969, after a decade of development work on centrifuges, the Japan Atomic Energy Commission started a three-year project to compare gaseous diffusion with centrifuge technology as alternative options for future Japanese enrichment development. In 1971 the Commission reached the conclusion that centrifuge technology would be chosen for further development leading to a Japanese uranium enrichment plant [39d]. In 1977 Japan's Power Reactor and Nuclear Fuel Development Corporation (PNC) began construction of a pilot centrifuge plant at Ningyo Pass (see figure 8.2) which is scheduled to reach its full capacity of about 70 t SWU/yr in 1981. This is 40 per cent higher than the original estimates because of technological improvements [39d]. A commercial gas centrifuge enrichment plant of 1 000−2 000 t SWU/yr is planned for the beginning of the 1990s. The commercial plant may be preceded by a demonstration plant of 250 t SWU/yr in 1985 [39e]. A commercial facility capable of supplying one-third of Japanese requirements is considered sufficient to mitigate the effects of any political or economic changes which could affect the reliability of foreign supplies. Finally, Japan is also involved in relatively intense research and development on laser [91] and chemical-exchange techniques (see chapter 6).

Figure 8.2. The Japanese centrifuge plant at Ningyo Pass

Japan has stated its willingness to "accept any kind of effective safeguard procedure or international custody agreed upon internationally" [39f] for preventing the proliferation of sensitive nuclear materials and technology. With respect to its own plans for a commercial enrichment plant Japan is "not very enthusiastic in promoting sales of uranium enrichment services . . . to other countries at this stage, without having an established supply scheme of uranium enrichment services agreed upon" [39g].

Japan favours a multinational uranium enrichment plant, managed internationally. It is ready to collaborate with other countries such as Australia in a joint enterprise. The preference for Australia is understandable because Australia has large uranium resources, which Japan lacks, and because Australian non-proliferation policies are generally considered to be very strict. The Japanese and Australian governments have already completed a two-year study of the feasibility of a joint enrichment project. The final report concluded in 1979 that Australia could construct the plant, employing centrifuge technology from several countries, including Japan [92].

South Africa

South Africa has very large uranium resources (391 000 t, not including those in Namibia) which it extracts as part of its gold-mining operations, and is one of the world's largest producers and exporters of uranium. Just under 4 000 t were produced in 1978 [41], and South Africa is the third (after the USA and Canada) major exporter of natural uranium in the world outside the centrally planned economies (see table 8.2). South Africa operates two research reactors [30b]. Two pressurized water power reactors with a total power output of 1.85 GW(e) are under construction at Koeberg by a French consortium [40]. South Africa is interested in marketing enriched uranium. An indigenous enrichment facility could also make the country independent of overseas sources of nuclear fuel for its future power reactors. Its current dependence on US enrichment services makes South Africa vulnerable to US pressure to sign the NPT, which South Africa has so far declined to do.

The Enrichment Corporation of South Africa Ltd (UCOR) operates a pilot enrichment plant of unspecified capacity at Valindaba. This facility has presumably begun to employ the Helikon technique described in chapter 6 (see figure 8.3). In 1975 plans were announced for a plant based on the same process with a capacity of 5 000 t SWU/yr to begin operation in 1986 [93]. However, in the late 1970s the South African government announced an alternative plan to enlarge the plant at Valindaba in order to fulfil its domestic needs for enriched uranium [94]. Given the size of the South African nuclear power programme, this would indicate a capacity of 200–300 t SWU/yr. A recent report states that a 300 t SWU/yr plant might start operation in 1983 [48c].

South Africa has not signed the NPT, but it has stated that it shares the objectives of the Treaty. Its objections have been based on possible

Table 8.2. The world enrichment picture

(Present involvement of countries in uranium enrichment in relation to their nuclear power programmes, uranium resources and adherence to arms control treaties).

Country	Enrichment status* — Installed capacity (t SWU/yr)[a] as of January 1981	Enrichment status* — Research programmes[b]	Enrichment needs for power reactors (t SWU/yr) in 1980[c]	Uranium resources — Reasonably assured (10^3 t)[d]	Uranium resources — Production 1978 (t/yr)[e]	Status with respect to arms control treaties[f] — NPT[g]	PTBT[h]	Treaty of Tlatelolco[i]
USA	25 300[j] (GD)	GD,GC,AE,L,CE,PL	6 000	708	14 200	R	R	PI:S PII:R
USSR	7 000–10 000[l] (GD)	GD,GC(?),L,CE(?)	600–1 100	?	7 000[k]	R	R	PII:R
Eurodif countries	6 000[l] (GD)							
France	300–600[m] (GD)	GD,GC,L,CE,PL	1 700	55.3	2 183			PI:S PII:R
	(3 010)[n]							
Italy	(1 050)[n]	GD,GC,L	140	1.2		R	R	
Belgium	(670)[n]	L	175			R	R	
Spain	(670)[n]		70	9.8	191	R	R	
Iran	(600)[n]					R		
Urenco countries								
UK	458[o] (GC)	GD,GC,L	240	4.5	41	R	R	PI:R PII:R
	400[p] (GD)							
	(213)[q]							
FRG	(245)[r] (GD)	GC,AE,L,PL	1 000			R	R	PI:R PII:R
		GC,PL						
Netherlands	180 (GC)	GD,GC,L	55			R	R	
China		GD,GC,L,CE		>50[s]	?	R		
Japan	30 (AE)	AE	1 600	7.7	2	R	R	
South Africa	6	AE		391	3 961		R	R
Brazil	0[t]	GC		74.2		S	R	R
Pakistan	?					S	R	
India	?	GC(?),L(?)	50	29.8	126			s
Argentina		GC,L,CE,PL		28.1		R		s
Australia	0[u]			299	516		R	
Israel		L						
Canada			30	235	6 803	R	R	
Sweden		Stopped	425	301		R	S	
South Korea		Stopped	60	4.4		R	R	
Taiwan			140			R	R	

Switzerland				R R
Namibia	133	2 697		R
Niger	160	2 060		R
Gabon	37	1 022		R
Zaire	1.8	0^v		S
Algeria	28			R
Central African Republic	18			R
Somalia	6.6			S

* Abbreviations of processes: GD = Gaseous Diffusion; GC = Gas Centrifuge; AE = Aerodynamic; L = Laser separation; CE = Chemical Exchange; PL = Plasma separation (electromagnetic processes).

a Included are all known enrichment plants; see country descriptions and appendix 8B.

b Many of these programmes are mentioned in the country studies; further descriptions can be found in. (a) INFCE, see reference [35b]; (b) Interdevelopment, see reference [48]; (c) Blumkin, see reference [54]; (d) Wilcox, see reference [124]. A question mark by a process indicates that there have been occasional reports about R&D activities. The present status of these programmes, however, is unknown.

c See appendix 8B; the figures here are rounded off.

d Resources which are expected to be exploited at costs lower than US$ 130/kg U; see reference [41], p. 18.

e See reference [41], p. 22.

f S = Signed; R = Signed and Ratified; see reference [97a].

g Treaty on the non-proliferation of nuclear weapons (Non-Proliferation Treaty–NPT).

h Treaty banning nuclear weapon tests in the atmosphere, in outer space and under water (Partial Test Ban Treaty–PTBT).

i Treaty for the prohibition of nuclear weapons in Latin America (Treaty of Tlatelolco).

Two additional Protocols are annexed to the Treaty, referred to in the table as PI and PII.

PI : Under *Additional Protocol I* the extra-continental or continental states which, de jure or de facto, are internationally responsible for territories lying within the limits of the geographical zone established by the Treaty (France, the Netherlands, the UK and the USA), undertake to apply the statute of military denuclearization, as defined in the Treaty, to such territories.

PII: Under *Additional Protocol II* the nuclear weapon states undertake to respect the statute of military denuclearization of Latin America, as defined in the Treaty, and not to contribute to acts involving a violation of the Treaty, nor to use or threaten to use nuclear weapons against the parties to the Treaty.

j Partly completed capacity; full capacity of 27 300 t SWU/yr is scheduled to be reached in 1983.

k See reference [45a].

l Partly completed capacity; full capacity of 10 800 t SWU/yr is scheduled to be reached in 1982.

m Gaseous diffusion plant at Pierrelatte.

n The total capacity of Eurodif is divided among the participating countries in proportion to their share in the Eurodif organization.

o Partly completed capacity; full capacity of 2 000 t SWU/yr is scheduled to be reached in the late 1980s.

p Scheduled to be closed down around 1985.

q Size of the Urenco capacity (pilot and demonstration plants) at Capenhurst, UK [39b].

r Size of the Urenco capacity (pilot and demonstrations plants) at Almelo, the Netherlands; this includes a 200 t SWU/yr joint West German–Dutch demonstration plant [39b].

s See country study of China.

t A capacity of 200–300 t SWU/yr is scheduled to start operating in the mid-1980s.

u Seriously considering construction of a commercial enrichment plant for export purposes.

v Production before 1975 was 25 600 t of uranium (reference [41], p. 22). At the beginning of the nuclear age, Zaire (Belgian Congo) had the main known uranium deposits.

risks of industrial espionage and the extra costs of the safeguards system which, in its view, may damage the competitive position of South African industry [55d]. South Africa must be regarded as a near-nuclear weapon state. There have been reports of preparations for a nuclear explosive test which were denied by the South African government [30c]. South Africa is politically isolated in the international arena because of its racial policies. Racial conflicts in South Africa are growing and many observers consider it to be a region with a serious risk of a major war. Given the persistent belief that nuclear weapons can play either a deterrent or combat role in a nation's military arsenal, it would not be surprising to find South Africa developing such weapons.

Figure 8.3. Helikon separation units at Valindaba
View of one of the stages of the pilot plant.

III. Projected facilities and R&D programmes

Brazil/Argentina

Brazil's nuclear energy programme has been motivated by anticipation of the future need of additional energy and the desire to diversify its reliance on energy resources. The ultimate goal is a complete domestic nuclear fuel cycle. A major step towards this goal was taken in 1975 with the conclusion of a 'nuclear package deal' with FR Germany. This agreement includes the supply by FR Germany of four 1.3 GW(e) LWRs, an option on four others, a pilot fuel fabrication plant, a pilot reprocessing plant and a uranium enrichment facility based on the jet nozzle process [76a]. In addition, Brazil has recently purchased a UF_6 conversion plant from France [79b]. In January 1981 two LWRs with a total capacity of 1.87 GW(e) (one supplied by the USA and the other as part of the West German sale) were under construction [40].

Brazil has assured uranium resources of about 75 000 tonnes [41]. As part of its resource development policy the Brazilian government announced plans in the mid-1970s to supply enrichment services to other countries and to construct a 2 000−2 500 t SWU/yr enrichment facility [95]. Because demand for enrichment services did not develop as rapidly as expected at that time, present plans include only a demonstration-size facility with a capacity of 200−300 t SWU/yr, on which construction is scheduled to start in 1982. Meanwhile, further development work is being carried out in a West German–Brazilian co-operation (which makes Brazil co-owner of the jet nozzle process) on a 24-stage experimental cascade with a capacity of 6.5 t SWU/yr now under construction at Resende (see figure 8.1) [96]. In addition to large uranium resources Brazil also has large thorium resources and is conducting a small R&D programme on a gas-cooled breeder reactor designed to breed ^{233}U out of thorium [56b].

The need for a major nuclear energy development programme in Brazil has been questioned by both foreign and domestic critics. In particular it has been suggested that Brazil's great hydropower resources could be developed at costs substantially lower than the sum it will have to pay FR Germany for the package deal [46b].

Brazil has not signed the NPT, offering as one reason that the Treaty discriminates against non-nuclear weapon states. An important aspect of Brazil's posture with respect to nuclear explosives is that it wants to keep open the option for carrying out nuclear explosions for peaceful purposes. Brazil has ratified the Treaty of Tlatelolco, but it has not waived the requirements laid down in Article 28 of the Treaty for it to take effect. One of these requirements is that the nuclear weapon states ratify Protocol II to the Treaty. The Soviet Union did so with the provision that it would consider the carrying out by any party to the Treaty of explosions of nuclear devices for peaceful purposes as a violation of its obligations under Article 1, and that this would be incompatible with its non-nuclear status. Because not all the requirements under Article 28 have been met, the Treaty is not yet in force for Brazil [97a].

Brazilian attitudes towards proliferation are affected to a substantial degree by the attitudes and actions of Argentina. Although Argentina has essentially no need for enrichment services, having opted for a nuclear economy based on natural uranium and heavy-water, Argentine positions on non-proliferation must be examined. Like Brazil, Argentina has cited the discriminatory nature of the NPT as its reason for refusing to sign the

Treaty [55a]. Also like Brazil, Argentina wants to keep open the option of using nuclear explosives for peaceful purposes. This desire was made explicit when Argentina signed the Treaty of Tlatelolco. So far, however, Argentina has not ratified this Treaty.

Brazilian–Argentine relations are often discussed in terms of regional rivalry, with both countries seeking technical and political leadership in South America. Although this rivalry cannot be excluded as a motive for both countries to seek the construction of nuclear explosive devices, the attitudes of the USA and USSR, who have not succeeded in stopping vertical proliferation, also play an important role. Both Brazil and Argentina see the NPT as a Treaty which puts severe constraints on the nuclear activities of non-nuclear weapon states, while the nuclear weapon states are hardly restricted in their activities. Both countries seek independence for their nuclear activities and in May 1980 signed an agreement on "peaceful cooperation in the research and development of peaceful nuclear energy applications". Three other accords detail specific co-operation in the fields of basic and applied research, uranium mining and the design, construction and operation of nuclear power plants [98].

Pakistan

Pakistan has been carrying out nuclear research for more than 20 years. It operates a 5 MW(th) (thermal) research reactor, supplied by the USA in 1965 [56c], and a 125 MW(e) heavy-water reactor supplied by Canada and operating since 1972 [40]. A study by the IAEA in 1975 concluded that Pakistan had an urgent need for more energy. According to its UN Ambassador, Pakistan had plans in 1976 to build 24 medium-sized nuclear power plants by the end of the century. He also stated that "fuel fabrication and reprocessing facilities and a heavy-water plant are ancillary to the plan and will be established as the programme is put into effect" [56d]. Earlier in 1976 Pakistan made a deal with France for a plutonium reprocessing plant, and the two governments have signed a trilateral safeguards agreement with the IAEA. The United States protested against this plan and put heavy pressure on both Pakistan and France to cancel the project. Construction of the project had already begun when France finally decided to terminate its participation and in 1979 the last French technicians left Pakistan [56a, 99]. By the end of 1980 Pakistan was reported to have continued construction of the plant independently, though the size of the plant would be "not much more than one-tenth of the size of the commercial reprocessing plant France had agreed to sell . . . " [129].

With respect to uranium enrichment it was revealed in 1978 that Pakistan is secretly building a gas centrifuge enrichment plant at Kahuta, near Rawalpindi. The plans for this facility were clandestinely obtained from the Dutch centrifuge plant in Almelo, and components and materials have been purchased in a number of Western countries, including FR Germany, the Netherlands, Switzerland and the USA [100,101]. The Pakistani facility is reported to be secured against attack with anti-aircraft missiles. Pakistan also operates a pilot gas centrifuge facility at Sihala [102].

Pakistan has hardly any indigenous uranium resources and is dependent upon imported nuclear fuel. Recently it has bought approximately 650 t of uranium from Niger via Libya [103].

Although Pakistan has not signed the NPT, it has submitted its operating nuclear facilities to IAEA safeguards [56e]. In the early 1970s the Pakistani government expressed positive views on the NPT, but it has consistently stated that it is unable to join the Treaty because of its apprehensions about Indian nuclear policies (see India below) [55e].

Pakistan denies that it is making a nuclear bomb and has advocated a nuclear weapon-free zone in southeast Asia [104]. Some analysts believe that even before the Indian nuclear explosion in 1974 Pakistan intended to manufacture a nuclear explosive [105a]. The perceived 'nuclear threat' from India and the wish to acquire an 'Islamic bomb' (supported by Libya) are often put forward as motivations for Pakistan to manufacture nuclear explosives. Pakistan's present work on the construction of a gas centrifuge enrichment plant is now widely regarded as aimed at the production of highly enriched uranium suited for the explosion of a nuclear device [62].

India

India has a long tradition in nuclear research, beginning even before the establishment of the Indian Atomic Energy Commission in 1948. A nuclear research reactor in Bombay was started in 1956 [56f]. India has always reserved its right to pursue an independent policy of exploiting nuclear energy for economic development and for the attainment of a maximum degree of economic self-reliance. In 1981 India operated 400 MW(e) of LWR capacity supplied by the USA and a 202 MW(e) HWR supplied by Canada. A capacity of 1.08 GW(e) of HWRs is under construction and scheduled for operation in 1984, and these reactors are all being manufactured by domestic Indian companies [40]. India is also building heavy-water production facilities which had been scheduled to begin operation in the near future [60]. However, these projects have been delayed by difficulties in providing sufficient electric power to operate them.

India has substantial uranium resources, estimated at some 30 000 tonnes [41]. By choosing to manufacture its own heavy-water reactors India has decided to develop a national fuel cycle based on natural uranium, but enrichment services are still required for its 400 MW(e) of LWRs, which need about 50 t SWU/yr. However, the Indian government has recently expressed a serious interest in uranium enrichment, primarily for national security reasons (see below). It has been reported that India might be carrying out some research on laser enrichment [48d, 54b]. There have also been reports that the Indian AEC set up a study programme on centrifuges in 1972, but no reports are known to have appeared since then [54c, 124]. In addition to its uranium resources, India also has large thorium resources and conducts, with French co-operation, research on thorium–uranium-233 breeder reactors [105b]. In the long term India's main interest lies in the development of fast-breeder reactors. In this connection research is being conducted on fuel reprocessing, and India has two small research facilities devoted to this effort [60].

In 1974 India demonstrated that it can manufacture and test nuclear explosive devices. This development was made possible by plutonium obtained from the 40 MW(th) heavy-water research reactor CIRUS, supplied by Canada [106]. The heavy water for the reactor came from the USA. The Indian government called its test a 'peaceful nuclear explosion' and justified it on economic grounds, especially for the development of its resources [56g]. However, this explosion qualifies India as at least a near-nuclear weapon state. India is not a party to the NPT, and has consistently been one of the most fervent objectors to the Treaty. The Indian government feels that the NPT "discriminates against non-nuclear states and denies them the benefits of peaceful nuclear technology and does nothing to further the cause of international disarmament" [56h].

In 1977 the Indian Prime Minister stated that India would not acquire nuclear weapons and that it would not conduct any more peaceful nuclear explosions [107]. However, in 1979 it was stated that if Pakistan persisted in its efforts to make a nuclear bomb, India would be forced to reconsider its earlier decision [97b]. This policy was reaffirmed in 1980 when Prime Minister Gandhi assured Parliament that there is no need for the country to feel any sense of insecurity about Pakistan's attempts to manufacture weapon-grade uranium. "While we do not have all that we would like to have in the defense sphere, we are trying to strengthen ourselves. So far as enrichment is concerned, . . . we want to be ready", she said, adding, "I do not think it would be proper to mention the details [of the experiments] publicly" [108]. However, in July 1981, the Prime Minister indicated that India would not manufacture nuclear weapons even if Pakistan did so [131], and that the policy of the government of India continues to be the utilization of atomic energy for peaceful purposes [132, 133, 134].

After the 1974 Indian nuclear explosion the main suppliers of nuclear materials to India demanded more guarantees against diverting nuclear materials from the civilian fuel cycle. Because India would not agree to full-scope safeguards (see chapter 3), Canada suspended and, in 1976, ended its supply of fuel and materials to India [56i]. After the Canadian withdrawal, the Soviet Union agreed to sell heavy-water to India but forced India to accept IAEA safeguards on completed and nearly completed heavy-water reactors [56j, 45e]. The United States suspended shipments of low-enriched uranium in 1974, but the embargo was lifted the

[56i]. The USA has tried to use its position as the main supplier of nuclear fuel to India to get India to accept full-scope safeguards on its nuclear facilities. Each US export licence for shipment of low-enriched and highly enriched uranium (for research reactors) to India is scrutinized by the US Nuclear Regulatory Commission to assure that no reprocessing of uranium will take place. However, this kind of unilateral pressure suffers from severe weaknesses as a proliferation control mechanism (see chapter 3).

Australia

Australia has no current plans for a domestic nuclear power programme, but does possess large uranium resources, estimated at 300 000 tonnes [41]. After a long political debate it was decided in 1977 that Australia should exploit these resources because of a "moral obligation to help other countries in assuring their energy supply and in order to stop the introduction of the plutonium technology" [109]. This decision was followed in 1979 by the announcement of the Australian government that it intended to study the possible development of a uranium enrichment industry in Australia. Consistent with this policy Australia has sought to export uranium in the highest possible upgraded form, including enriched. This desire is reflected in the most recently proposed sales contracts in which the buyer is obliged (upon eight years advance notification) to accept 35 per cent of the contracted uranium in enriched form [79a].

Research on enrichment, in particular the gas centrifuge and laser techniques, has been conducted in Australia since 1965 [35j], and in 1973 the Australian Atomic Energy Commission became a member of the so-called Association for Centrifuge Enrichment (ACE) initiated by Urenco. The ACE studied various aspects of centrifuge plant usage, including technology, construction and finance [110]. Since 1973 Australia has had contacts with a number of countries, including France, Japan and the United States, in addition to the Urenco countries, to explore possible collaboration or purchase of technology [79a].[2] Australia is also conducting research on chemical exchange and plasma enrichment [48d].

Australia is party to the NPT. It is not a member of the Nuclear Suppliers Group (see chapter 7), but it has stated that it will apply export criteria which satisfy the Group's guidelines [39h]. Australian export of uranium will only take place if the following conditions are satisfied in bilateral agreements [79a]:

1. The importing country, if not a nuclear weapon state, must have signed the NPT, so that the supplied material is subject to international control by the IAEA.

---

[2] As this book was going to print, Australia announced it had selected Urenco for a joint venture to help Australia build a centrifuge enrichment plant and develop a uranium enrichment centrifuge [146].

2. Uranium supplied by Australia is not allowed to be enriched to a higher concentration than 20 per cent without Australian consent.

3. Australian consent is also required for the reprocessing of any uranium fuel supplied by Australia.[3]

Bilateral agreements under these conditions have been concluded so far with Finland, France, South Korea, the Philippines, the UK and the USA [79a].

Israel

Israel is a technologically advanced country and has had a nuclear research programme for several decades. It has no nuclear power plants but is considering the construction of one in collaboration with the United States. Israel does have two research reactors, one of 5 MW(th) fuelled with highly enriched uranium and the other a 26 MW(th) heavy-water reactor [56k]. In this connection Israel has also constructed a pilot heavy-water plant and also possesses a small fuel-reprocessing facility at Dimona [60]. This combination of facilities gives Israel the capability of producing enough plutonium for at least one nuclear weapon a year using only its domestic installations.

Israel has only minor uranium resources, but is reported to fuel its 26 MW(th) heavy-water research reactor at Dimona with uranium extracted from phosphate resources [55f]. It is also believed that the Israelis clandestinely acquired 200 tonnes of uranium loaded on a ship registered to a different purchaser [111]. Research on laser enrichment has been conducted in Israel for many years [55a]. However, Israel is not known to have a uranium enrichment facility of any significant capacity.

Israel has not signed the NPT, citing reasons of national security. Although official Israeli policy has remained rather ambiguous about nuclear weapons, a knowledgeable Israeli spokesman has stated that Israel is capable of assembling nuclear explosive devices in a short time [112]. Such a capability is considered essential by the Israelis to provide a deterrent against attacks by Arab countries. Other reports state that at least since the 1973 Yom Kippur War Israel has acquired between 10 and 20 nuclear weapons or can assemble them in a very short period of time [3d, 113].

---

[3] These conditions do not completely coincide with the Nuclear Suppliers Group guidelines.

Table 8.3. The projected world enrichment picture

Countries	1985 Enrichment capacity (t SWU/yr)	1985 Enrichment needs for power reactors (t SWU/yr)[a]	1985 Attainable uranium production (t/yr)[b]	1990 Enrichment capacity (t SWU/yr)	1990 Enrichment needs for power reactors (t SWU/yr)[a]	1990 Attainable uranium production (t/yr)[b]
USA	27 300	12 500	34 100	29 500	15 200	44 200
USSR	7 000–10 000	>930[f]	7 000[g]	7 000–10 000	>930[f]	7 000[g]
Eurodif countries	10 800			10 800		
France	300–600[c] 50–100[d] (5 430)[e]	4 700	4 020	300–600[c] 50–100[d] (5 430)[e]	5 450	4 020
Italy	(1 890)[e]	600	120	(1 890)[e]	360	120
Belgium	(1 200)[e]	600		(1 200)[e]	600	
Spain	(1 200)[e]	860	1 272	(1 200)[e]	1 400	1 272
Iran	(1 080)[e]			(1 080)[e]		
Urenco countries	1 600[h]			2 000[h]		
UK	(400–700)	600		(700)	850	
Netherlands	(500–1 000)	55		(900–1 300)	55	
FR Germany	(0–400)	2 100	200	(0–400)	2 250	200
China	180	?	?	180	?	?
Japan	70–250	3 100	30	250–2 000	2 650	30
South Africa	200–300	200	10 600	200–300	200	10 400
Brazil	200–300	65	970	200–300	325	970
Argentina			680			680
Pakistan	?			?		
India	?	50	200	?	50	200
Australia	?		12 000	?		20 000
Canada		30	14 400		30	15 500
Sweden		1 300	400		1 100	400
South Korea		500			800	
Taiwan		700			550	
Switzerland		320			425	
Niger			10 500			12 000
Namibia			5 000			5 000
Gabon			1 500			1 500
Algeria			1 000[i]			1 000[i]
Central African Republic			1 000			1 000

[a] See appendix 8B; the figures here are rounded off.
[b] See reference [41].
[c] Gaseous diffusion plant at Pierrelatte.
[d] Chemical-exchange pilot plant.
[e] The total capacity of Eurodif is divided among the participating countries in proportion to their share in Eurodif, assuming that the division of shares will not change.
[f] Very little public information is available on Soviet plans for nuclear power growth. The minimum given here refers to the power plants which are scheduled to start operation in the early 1980s.
[g] No data are available about future plans. The given 7 000 t/yr refers to production in the late 1970s [45a].
[h] As of January 1981 the division of plants among the three participating countries was not yet completely agreed upon. However, it was known that the capacity in Almelo will include in part a joint West German–Dutch plant.
[i] See reference [123].

IV. Substantial enrichment needs or uranium resources

Canada

Canada's uranium resources are estimated at 235 000 tonnes, and it is one of the largest uranium producers in the world (6 800 tonnes in 1978) [41]. Canada is a major exporter of both nuclear materials and technology and intends to play a significant role in this field. The dominant reactor type is the heavy-water reactor (CANDU) which is fuelled by natural uranium [40]. Thus, uranium enrichment does not play an important role in this programme. In the mid-1970s there was a brief co-operation between France and Canada in the enrichment field (Canadif). However, the results of a pilot study for a 9 000 t SWU/yr gaseous diffusion plant near Quebec raised doubts in Canada as to whether such a plant would be a wise investment. A plant with such a large capacity would consume most of the domestic uranium production, and the stagnating world market for enrichment services gave no promise that enough contracts could be obtained to justify such a capacity [76b].

Canada has a considerable indigenous nuclear power programme. By the end of 1980, 10 power reactors with a power output of 5.5 GW(e) were operating, and 14 power reactors with an output of 9.8 GW(e) were planned or under construction [40]. A major constraint on Canada's uranium export policy is the desire to assure domestic supplies for the next 30 years. Approximately 20 per cent of the present known uranium resources have to remain in the country, and foreign investments in uranium production facilities are limited to 33 per cent. Foreign supply contracts have a maximum term of 10 years [114]. Another constraint on exports is enforced by Canada's non-proliferation policy, which became much more stringent after the Indian nuclear test in 1974. This event made clear that the Canadian-supplied research reactor which produced the plutonium for the device was not adequately safeguarded [56e]. Canada now requires for contracts with non-nuclear weapon states a binding commitment to non-proliferation and international safeguards on *all* nuclear activities, current and future, indigenous or imported [47b].

Sweden

In a 1980 referendum 58 per cent of the Swedish population voted in favour of pursuing Sweden's current nuclear power programme. At the end of 1980 Sweden had a capacity of 3.7 GW(e) in operation with an additional capacity of 5.7 GW(e) under construction and scheduled to be completed by 1985 [40]. This is considerably lower than earlier Swedish projections which had anticipated an installed capacity of 23 GW(e) by 1985 [115]. When these projections were made in the early 1970s, Sweden also attempted to diversify its source of enriched uranium by acquiring a 10 per cent share in the Eurodif consortium. Some research was also undertaken on centrifuges [54c]. However, in 1974 Sweden had already contracted with the USA and the USSR for enrichment services sufficient for its needs of the time, and as expansion of its nuclear programme began to appear more uncertain, Sweden withdrew from the industrial phase of Eurodif [116]. Research on centrifuges was abandoned in 1977 [54c].

Sweden has large uranium resources, estimated at about 300 000 t. However, these are difficult to mill, and the country's production will probably not amount to more than 400 t/yr by 1985 [41].

Sweden has signed and ratified the NPT.

South Korea

South Korea is a rapidly industrializing country, and to reduce its dependence on imported oil it has embarked on a large nuclear power programme. Since 1978 one LWR with a capacity of 564 MW(e) has begun operation, and LWRs with a total power output of 6.2 GW(e) supplied by France and the United States are under construction. Furthermore, a 629 MW(e) Canadian-supplied HWR is scheduled for completion in 1982 [40]. South Korea is aiming at fulfilling 40 per cent of its energy needs by nuclear power in 1991 [117]. Enrichment services are provided by the USA [57a], except for two recently ordered power plants which are to be supplied and fuelled by France for a period of 10 years, starting in 1986 [117]. South Korea has very small indigenous uranium resources (4 400 tonnes [41]) and no enrichment facilities or substantial R&D programmes in this field. However, France is reported to have considered the possibility of building a uranium enrichment plant in South Korea [117]. The South Koreans were also interested in acquiring a French reprocessing plant, but under US pressure this deal has been cancelled [29a, 56m].

South Korea ratified the NPT in April 1975. However, their reactions in the mid-1970s to the possible withdrawal of the nuclear weapons accompanying US troops stationed in South Korea indicate the limits of the South Korean willingness to renounce nuclear weapons. Two months after ratification of the NPT, South Korean President Park stated that South Korea would do everything in its power to defend its national security including the manufacture of nuclear arms if the US nuclear umbrella were to be withdrawn, a contingency he regarded as unlikely [56m]. A similar statement was made by the South Korean Minister of Foreign Affairs in June 1977: "We have signed the NPT, and thus our basic position is that we do not intend to develop weapons by ourselves. But if it is necessary for national security interests and people's safety, it is possible for Korea as a sovereign state to make its own judgement on the matter" [118].

Taiwan

Taiwan is a rapidly industrializing country with relatively sophisticated domestic technology, including research and development programmes at two nuclear research centres. It has embarked on a rather large nuclear power programme which consisted in 1981 of two operating LWRs with a total power output of 1.21 GW(e). Another 3.72 GW(e) of LWR capacity is now under construction and scheduled for operation by 1985 [40].

Taiwan has no indigenous enrichment facilities and its present and future enrichment needs are covered by US contracts [57]. It has carried out research on reprocessing, but when in the early 1970s US inspectors found that Taiwan had built a reprocessing laboratory facility, the US government put pressure on Taiwan to stop reprocessing nuclear materials [56n]. Plans for obtaining a reprocessing facility from France were also cancelled under US pressure [29a].

Taiwan has no known uranium resources, so it is dependent on foreign imports of both uranium and enrichment services. The major source is the United States. Taiwan ratified the NPT in 1970. This was consistent with its declaratory policy of pursuing only civilian uses of nuclear energy and nuclear research generally. However, Taiwan's expulsion from the United Nations in 1971 isolated it internationally. Many states switched their diplomatic and economic ties to the People's Republic of China, and Taiwan was also ousted from the IAEA in 1972. With US encouragement, however, Taiwan has continued to accept the full range of IAEA safeguards, including on-site inspection of all its nuclear facilities and materials [56o]. In the past Taiwan has considered the acquisition of nuclear weapons. In 1975 the Taiwanese Premier stated that 17 years before, Taiwan had begun research on nuclear weapons and considered acquiring a nuclear arsenal, but the idea was finally rejected by President Chiang Kai Shek [56n].

Switzerland

Switzerland's nuclear power generating capacity in 1980 was about 1.9 GW(e), and this is scheduled to be doubled by the end of the 1980s [40]. All of these reactors are light-water moderated and thus use slightly enriched uranium as fuel. Most of the needed enrichment services will be supplied under contract with the United States [57a].

Switzerland is not known to have plans for domestic uranium enrichment or fuel reprocessing. Present reprocessing of Swiss spent fuel is done by France and the UK [44j].

The Swiss have ratified the NPT and have concluded a safeguards agreement with the IAEA. Switzerland has also participated in the Nuclear Suppliers Group (see chapter 7) and is an active exporter of nuclear technology. For example, Argentina has recently purchased a heavy-water plant from Switzerland [119].

African countries

The known African uranium resources are concentrated in the following countries (other than South Africa): Namibia, Niger, Gabon, Zaire, Algeria, Central African Republic and Somalia. Exploration for uranium is going on in a number of other countries as well. Although not highly industrialized nor technologically advanced and at present not known to have nuclear power programmes nor to plan the construction of indigenous enrichment facilities, these countries are included in this survey because they could be potential candidates for the location of an enrichment plant through co-operation with industrialized countries. Such co-operation can provide advanced technology to developing countries in exchange for uranium supplies.

Namibia's uranium resources are currently estimated to be 133 000 t [41]. Several foreign firms, including mining companies from Canada, France, South Africa and the UK, have begun exploration for the development of Namibian ore [77a]. The Rössing mine, operated by the London-based Rio Tinto Zinc firm, is the most important production facility. The 1974 United Nations Declaration on the Natural Resources of Namibia makes illegal any exploitation of Namibia's wealth for the benefit of South Africa [77b]. Production of Namibian uranium in 1978 was 2 700 tonnes; it could be increased to 5 000 tonnes by 1985 [41].

Two other important producers of uranium ore are Gabon and Niger, two former French colonies. France has close ties with these countries and has carried out most of the exploration, mining and marketing [52]. Niger's uranium resources are currently estimated at 160 000 tonnes [41], and the country's economy is heavily dependent on the mining of uranium. Niger is prepared to sell its uranium to any country willing to pay a fair price [120]. Recently Niger sold 450 tonnes of uranium to Libya, even though these two countries are generally considered to have rather hostile relations [120]. IAEA sources report that Libya probably resold the uranium to Pakistan and possibly other Arab countries [105a]. Niger is also reported to have sold 60 tonnes of uranium directly to Pakistan and 100 tonnes to Iraq [120]. Niger allows foreign firms to explore and mine uranium under negotiated conditions. Companies from France, FR Germany, Italy, the USA and Niger itself are participating in the exploitation [120,121]. The total production in 1978 was 2 060 tonnes and by 1985 annual production could exceed 10 000 tonnes [41]. Niger has not signed the NPT.

Gabon's uranium resources are currently estimated at approximately 37 000 tonnes [41]. Mining is carried out by France and by Gabon itself through the Comuf Company which is 75 per cent French and 25 per cent owned by Gabon [52]. In 1978 uranium production in Gabon was

1 022 tonnes and by 1985 a production of 1 500 tonnes might be achieved [41]. Gabon ratified the NPT in 1974.

In the mid-1970s Zaire was reported to be considering the construction of a 9 000 t SWU/yr gaseous diffusion plant [36c]. Such a plant would have been far too large to handle only Zaire's resources, which are estimated at only 1 800 t [41]. However, it was believed that the abundant supply of cheap hydropower in Zaire might make such a plant a profitable investment. Some studies of this idea were undertaken in co-operation with Belgium, but there have been no recent reports of progress on the project. Zaire ratified the NPT in 1970.

Algeria's uranium resources are currently estimated at 28 000 tonnes [122], and in 1980 plans were announced to exploit part of these resources. The construction of a mine and additional infrastructure are planned to start in 1981, and full production of 1 000 t/yr is scheduled to be reached in 1984/85 [123]. Algeria has no nuclear power plants but a 'prefeasibility' study is being undertaken for the generation of 10 per cent of Algeria's electricity needs by nuclear power plants by 1990 [123]. Algeria has not signed the NPT.

Uranium resources of the Central African Republic are estimated at 18 000 tonnes. There is no current exploitation of these resources, although the possibility exists that production could reach 1 000 t/yr by 1985 [41].

Somalia's resources are estimated at 6 600 tonnes, but there are no announced plans to develop these resources in the near future. Both the Central African Republic and Somalia ratified the NPT in 1970.

V. Concluding remarks

The most important data from the above national profiles are summarized in tables 8.2 and 8.3. They clearly show four major producers of enrichment services for the next decade: the USA, the USSR, and the two consortia, Eurodif and Urenco. Other countries which could become substantial contributors are South Africa, Brazil, Australia and Japan. A number of other countries could also develop relevant facilities, in particular Pakistan and India. As shown in chapter 6, gas centrifuge technology has become mature and is considered a first-rate option for new facilities. Meanwhile, research and development on laser enrichment techniques are being pursued actively in many countries.

An analysis of the enrichment market (see appendix 8B) reveals a substantial excess of supply over demand for the foreseeable future. Annual production capacity of separative work units by 1990 will exceed annual demand by 35 per cent, and cumulative production over the next decade will exceed cumulative demand by 40 per cent. This excess capacity coupled with the refusal of a number of the countries with enrichment capabilities to sign the Non-Proliferation Treaty generates considerable concern over the risks of future proliferation of nuclear weapons. These risks have been analysed in more detail in chapter 3.

Appendix 8A

Nuclear power growth 1980–1990

Estimates of future nuclear power growth can be found in several sources. A major problem in making such estimates is the presence of uncertain factors for which assumptions must be made. These factors include, for example, economic growth projections, the relation between economic growth and growth of energy consumption, the share of nuclear energy in the total energy supply, public acceptance of nuclear energy, and so forth. Recent history has shown that these factors can vary rapidly within relatively short periods of time, so long-term projections of nuclear power growth must, of necessity, have large margins of uncertainty. However, if projections are confined to a limited period of approximately 10 years, it is possible to make a reasonably reliable projection of the maximum capacity of nuclear power generation in the world outside the centrally planned economies area (WOCA). This is made possible by the time delay between the ordering of a nuclear power plant and the beginning of commercial operation. In most of the WOCA countries this delay is over 10 years; for example, there is a 12-year time delay in the USA [37b]; 10 years in FR Germany [135]; and 10–15 years in Japan [136]. Only in France is this period reported to be shorter [135]. Therefore, the great majority of nuclear power plants ordered in the 1980s will not come into operation until the 1990s.

From the point of view of adjusting enrichment capacity to projected needs, a 10-year projection of nuclear power capacity is sufficient because the construction time for additional enrichment capacity is relatively short. For a gaseous diffusion plant the construction time is approximately six years, for a jet-nozzle plant five years, and for a gas centrifuge plant four years [57b].

The main source that we used for estimating nuclear power growth is the "World list of nuclear power plants", published twice a year by the American Nuclear Society. This list specifies every nuclear power plant in the world of at least 30 MW(e) capacity which is in operation, under construction or on order and gives the actual or expected date of commercial operation. The criterion for listing a unit is that either an order or a letter of intent has been signed for the reactor. The figures used here are based on the list published in February 1981 [40], which gives the status as of 31 December 1980. Simply by ranking the listed reactors in order of their expected date of commercial operation, it is possible to obtain an estimate of the growth of nuclear power in the 1980s.

The expected date of commercial operation used in the list has been given by the utility or agency that owns the plant. Generally, owing to various delays the actual date of commercial operation is later than expected. For example, a comparison of the list published in August 1979 [137] with the one published in February 1981 [40] shows an average delay of about one year beyond the expected date of commercial operation. Furthermore, because some of the listed reactors will be retired in the 1980s, and because, as experience shows, some reactor orders can be expected to be cancelled, our estimate of nuclear power growth in the 1980s will give an upper limit for the attainable capacity in succeeding years.

The results of the projection are given in tables 8A.1 and 8A.2 and in figure 8A.1.

In the tables, a distinction is made between the World Outside the Centrally Planned Economies Area (WOCA) and the Centrally Planned Economies countries (CPE) in order to be able to compare the SIPRI estimates with those of other sources which usually consider only the WOCA. This separation is often made because information on both nuclear energy expansion plans and construction time delays is much harder to obtain from CPE countries. In both figure 8A.1 and tables 8A.1 and 8A.2 SIPRI estimates are compared with those of the International Fuel Cycle Evaluation (INFCE) committee which were published in February 1980 by the International Atomic Energy Agency [138].

Table 8A.1a. Nuclear power growth in the world outside the centrally planned economies area (WOCA) on a country-by-country basis (capacity in MW(e))[*]

Country	1980	1985	1990	Later/indefinite[a]
Argentina	335	935	935	692
Austria	692	692	692	
Belgium	1 650	5 450	5 450	
Brazil	0	626	3 116	
Canada	5 476	10 320	14 475	881
Egypt	0	0	622	
Finland	1 500	2 160	2 160	
France	12 818	41 008	52 268	3 900
FR Germany	8 576	17 657	21 274	6 284
India	602	1 684	1 684	
Iraq	0	0	0	900
Italy	1 387	3 391	3 391	1 904
Japan	14 552	24 034	24 334	
Korea, South	564	3 598	7 398	
Libya	0	0	0	300
Luxembourg	0	0	0	1 250
Mexico	0	1 308	1 308	
Netherlands	495	495	495	
Pakistan	125	125	125	
Philippines	0	620	620	
South Africa	0	1 844	1 844	
Spain	1 073	8 393	13 397	1 000
Sweden	3 700	9 410	9 410	
Switzerland	1 940	2 882	3 807	1 140
Taiwan	1 208	4 924	4 924	
Turkey	0	0	440	
UK	8 080	11 780	14 420	
USA	53 606	105 672	138 644	25 180
Total	**118 397**	**259 008**	**327 233**	**43 431**

[*] Figures for end of each of years 1980, 1985 and 1990.
[a] Scheduled to come into operation in the 1990s or at an unknown date. However, because actual construction in most cases has not yet started, the contribution of this category in the 1980s will not be significant.

Table 8A.1b. Nuclear power growth in the centrally planned economies area (CPE) on a country-by-country basis (capacity in MW(e))*

Country	1980	1985[a]	1990[a]	Later/indefinite[b]
Bulgaria	880	1 760	1 760	
Czechoslovakia	990	3 630	4 510	
German DR	2 270[c]	2 710	2 710	
Hungary	0	1 760	1 760	
Poland	0	440	880	
Romania	0	440	1 040	
USSR	20 395[c]	23 795	23 795	1 000
Yugoslavia	0	615	615	
Total	**24 535**	**35 150**	**37 070**	**1 000**

* Figures for end of years 1980, 1985 and 1990.
[a] Because neither detailed plans nor the 'construction time' of reactors is known, the figures for later years are less reliable as an estimate of nuclear power growth.
[b] Scheduled to come into operation in the 1990s or at an unknown date. However, because the actual construction in most cases has not yet started, the contribution of this category in the 1980s will not be significant.
[c] In the world list of nuclear power plants a capacity of 11 475 MW(e) (USSR) and 1 390 MW(e) (GDR) is reported to be completed [40]. Furthermore, a capacity of 8 920 MW(e) (USSR) and 880 MW(e) (GDR) is reported to be scheduled for completeion before 1981, though no present status is given. In this table the completed capacity and the capacity scheduled to be completed are added together.

Table 8A.2a. Nuclear power growth, 1980–1990, on a year-by-year basis in the world outside the centrally planned economies area (WOCA). Comparison of SIPRI estimates with INFCE estimates [138] (capacity in GW(e))*

	1980	1981	1982	1983	1984	1985	1986	1987	1988	1989	1990
SIPRI estimates	118	144	170	203	233	260	284	295	312	323	328
INFCE low	144	163	178	201	222	257	290	325	360	398	434
high	159	181	203	227	256	303	343	387	432	476	534

* Figures are for end of each year.

Table 8A.2b. Nuclear power growth, 1980–1990, on a year-by-year basis in the centrally planned economy countries. Comparison of SIPRI estimates and INFCE estimates [138] (capacity in GW(e))*

	1980	1981	1982	1983	1984	1985	1986	1987	1988	1989	1990
SIPRI estimates[a]	25	27	30	33	34	35	36	37	37	37	37
INFCE low	23	28	32	37	41	49	55	67	77	87	98
high	28	35	43	54	65	82	92	109	127	145	165

* Figures are for end of each year.
[a] Estimates for later years are less reliable, owing to lack of detailed information.

Figure 8A.1. Projection of nuclear electric capacity 1980–1990 in the world outside the centrally planned economics area (WOCA)

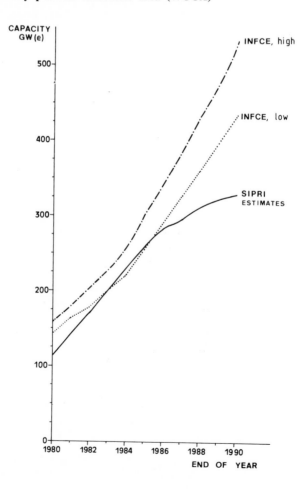

Appendix 8B

D∘mand for and supply of enrichment services

Demand for enrichment services

In order to be able to derive the future demand for enrichment services from the projected nuclear power growth presented in appendix 8A, assumptions must be made for several parameters involved. These include: (*a*) reactor type to be used, (*b*) load factors of the reactors, (*c*) operating tails assays at enrichment plants, and (*d*) amount of uranium and/or plutonium recycling.

The following estimates for these parameters are as realistic as possible, but in any case they are made in such a way that the calculated demand for enrichment services will most likely be an upper limit to the actual demand in the 1980s.

Reactor types

This estimate comes from the same source as that used to project nuclear power growth in the 1980s in appendix 8A [40]. These data show that in the near future about 95 per cent of all reactors will be light-water reactors (LWRs), which use slightly enriched uranium as fuel. The remaining 5 per cent will consist largely of heavy-water moderated reactors (HWRs), which use natural uranium as a fuel and thus have no need for uranium enrichment. Of the LWRs about one-third will consist of boiling water reactors (BWRs) while the remaining two-thirds will be pressurized water reactors (PWRs). Other current or possible future reactor types are the fast breeder reactor (FBR), the gas-cooled reactor (GCR) and the advanced gas-cooled reactor (AGR).

249

Reactor load factors

The load factor of a power plant is the ratio of the actual generated energy over a certain period of time to the energy the plant could have produced if it had been continuously operated at full capacity during that period. Operating experience with BWRs and PWRs (the dominant reactor types in the near future) shows that the average load factor for most reactors is between 50 and 70 per cent [139]. The US Office of Technology Assessment (OTA) gives the following formula for estimating the load factor as a function of time: 40 per cent in the first year, 65 per cent in the second year, followed by 12 years at 75 per cent. After the fourteenth year, the capacity factor drops by about 2 per cent every year to 35 per cent in the last (thirtieth) operating year [57c]. The Organization for Economic Cooperation and Development (OECD) assumes an average load factor of 70 per cent [95b]. Because most of the operating reactors in the 1980s will be less than 12 years old, the OTA conversion factors are used, which assume an overall load factor of 75 per cent for calculating the need for enrichment services [57c]. Using this high value will be consistent with the purpose of computing an upper limit for enrichment demand.

Operating tails assay

The tails assay of an enrichment plant is the percentage of ^{235}U in the depleted uranium. A higher tails assay means that less separative work but more natural uranium feed are required for a given amount of enriched product (see chapter 5, section III). Given the prices of natural uranium and enrichment services it is then possible to derive an optimum tails assay, one which minimizes the price per kilogram of enriched product. If the separation work price is denoted by C_s (in \$/SWU) and the price for natural uranium by C_f (in \$/kg U), then table 8B.1 gives an indication of the optimum tails assay as a function of the ratio of these prices (assuming 3 per cent enrichment of product) [140].

The price of natural uranium depends on when the contracts for deliveries are concluded. Prices under contracts negotiated in the period 1977 to mid-1979 have stabilized at a level slightly above \$104/kg U. Prices under contracts concluded earlier are considerably lower [145]. The US price for uranium enrichment in early 1980 was about \$100/SWU, which was reported to be some \$30–\$50 lower than the prices of the European suppliers Eurodif and Urenco [42a]. Considering table 8B.1, this means that from a commercial point of view, especially for the two European enrichment companies, the optimum tails assay will be well above 0.2 per cent. If the customer can choose the tails assay, which is the case for Urenco [141a] and Eurodif [141b], it is likely that a tails assay well above 0.2 per cent will be chosen. The US Department of Energy, however, has fixed its operating tails assay at 0.2 per cent, which is scheduled to be scheduled to be maintained throughout the 1980s [37c]. In the calculations an operating tails assay of 0.2 per cent is assumed, which is lower than the average that can be expected. This again leads to an overestimate of the potential demand for enrichment services in the 1980s.

Uranium and/or plutonium recycling

The percentage of ^{235}U in spent fuel from LWRs is usually around 0.9 per cent [57c], somewhat higher than the natural level of 0.71 per cent. Recycling of this uranium would reduce slightly the demand for enrichment services. Recycling of plutonium would imply the replacement of uranium by plutonium as a fuel and would therefore eliminate all need for enrichment of the replaced portion. However, it is highly unlikely that recycling of uranium and plutonium will take place on a substantial scale in the 1980s. Both the technical and political obstacles to such recycling remain substantial. Therefore, we will assume that no recycling of uranium or plutonium will take place in the next decade.

Summary of world enrichment demand

With these assumptions it is now possible to convert projected growth in nuclear electricity capacity to a projected demand for enrichment services in the 1980s. The relevant conversion factors for different reactor types are given in table 8B.2. Using these conversion factors the world demand for enrichment services through 1990 can be computed. The results are given in tables 8B.3 and 8B.7 and figure 8B.1. It must be noted that the results of appendix 8A cannot be used directly to derive the demand for enrichment services because appendix 8A does not take into account the various reactor types. A rough estimate can be obtained from appendix 8A by assuming that 62 per cent of all nuclear plants are PWRs, 30 per cent are BWRs, 3 per cent are AGRs, and 5 per cent do not need enrichment services. (This division holds only for the global distribution of reactors. For individual countries wide variations exist.) However, the results of tables 8B.3–8B.7 are based on the precise distribution of reactor types. The results in the tables and figure are only for the world outside the centrally planned economies areas (WOCA). This is the result of an attempt to evaluate the future market situation for enrichment services. In the centrally planned economies (CPE) no free market exists; the USSR is the sole supplier of enrichment services, and its supplies only meet the actual demands. Furthermore, the actual status of nuclear power generation in some CPE countries is uncertain (see table 8A.1b, p. 247). Depending on whether none or all of the 'uncertain' capacity is actually completed before 1981, the demand for enrichment services in CPE countries in 1980 would be somewhere between 968 and 1733 t SWU. For the Soviet Union alone the 1980 enrichment needs would be between 578 and 1132 t SWU. However, the USSR does supply some enrichment services to Western Europe, and these are included in the SIPRI analysis. Also included is Yugoslavia's demand for enrichment services, because this will be supplied by the United States [57a].

Figure 8B.1. Projected enrichment market 1980–1990

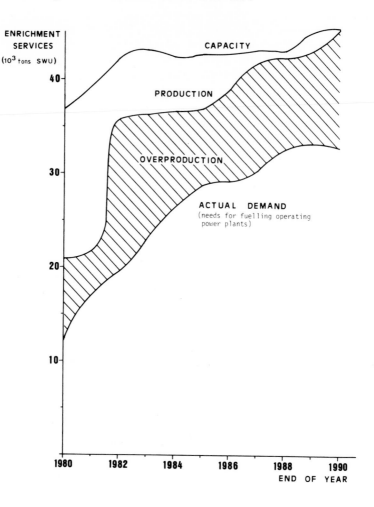

Supply of enrichment services

Table 8B.4 presents the projected development of uranium enrichment *capacity* in the 1980s. Some of the data can be found in the individual country studies in chapter 8, section II. For more detailed data some additional references are given in the table. Table 8B.4 gives only those capacities for which rather firm plans existed in the early 1980s. As shown in the country summaries, additional capacity in the late 1980s and early 1990s may be constructed by Australia, Japan and/or Coredif. With respect to the USSR, only the capacity which is available for non-CPE countries is included in table 8B.3. The Soviet figures presented here are taken from INFCE [143]. The US Department of Energy (DoE) has also estimated Soviet enrichment supplies [57d]. These estimates are on average about 25 per cent lower than the INFCE estimates. However, because the INFCE figures are based on data presented by the participating countries, these figures are used here. The difference between INFCE and DoE figures changes the global total by only a few per cent.

Table 8B.5 gives the expected *production* of uranium enrichment services in the 1980s. These data cannot be derived simply by adding all enrichment capacities, because the USA will not operate its enrichment plants at full capacity until the late 1980s in order to maintain production consistent with demand for US enrichment services [144]. With respect to the other enrichment suppliers it is assumed that their full capacity will be used. Furthermore, only the four major suppliers of enrichment services are included in table 8B.5, because production schemes of the others are uncertain and, in any event, their production will have only a marginal effect on global supply figures.

Table 8B.6 gives a summary of all non-US utilities contracts for enrichment services with the four major suppliers [57a].

Comparison of supply and demand

A more detailed look at the contracted supplies shows that roughly all production in the 1980s is contracted, assuming that the USA lowers its production to the level of the contracted demand. The Soviet Union's capacity available for the WOCA is equal to the contracted supplies. Eurodif production in the period 1980–1990 will be approximately 110 000 t SWU, while their contracted supplies for that period amount to 107 000 t SWU [57d]. Urenco contracts for the period 1980–1987 require deliveries of about 9 500 t SWU [39b], while their production during that period will be 8 750 t SWU (see table 8B.5). Demand is thus slightly larger than production in that period, but Urenco may be able to meet the demands because some production has already taken place in the late 1970s.

Therefore, the projected supply of enrichment services in the 1980s should very nearly equal the *contracted* demand. But, as will be shown below, the contractual obligations to purchase enrichment services substantially exceed the enrichment supplies necessary for the actual fuelling of nuclear power plants. In other words, many countries find themselves overcommitted to purchasing enrichment services, and this translates globally into a sizeable excess of supply over demand.

Table 8B.7 compares the actual demand of enrichment services, based on our projection of nuclear power growth in the 1980s, with the projected production of enrichment services during the same period. The table shows that by 1990 the cumulative production of enrichment services can be expected to exceed cumulative demand by 40 per cent. Annual production in 1990 will be about 35 per cent higher than the annual demand. In fact the situation is likely to be even worse than this, since these estimates are based on a conservative upper limit for enrichment demand. And if any more suppliers enter the market, the problem will be further aggravated. However, with the demand picture as bleak as it appears to be there will be strong economic pressures against the entry of new enrichment suppliers in the next decade.

Table 8B.1. Optimum tails assay for various ratios of feed to SWU prices for a 3 per cent enriched product

C_f/C_s	0.5	0.75	1.0	1.5	2.0
Optimum tails assay (%)	0.3	0.26	0.23	0.19	0.16

C_f = price for natural uranium (feed) in \$/kgU.
C_s = price for separative work in \$/SWU.

Table 8B.2. Demand for enrichment services (in tons SWU) per GW(e) rated capacity for three types of nuclear generating facility[*]

	Reactor type		
	PWR[a]	BWR[a]	AGR[a,b]
Initial loading	239	222	193
Annual replacement loading	105	117	96
Source	OTA [57c]	OTA [57c]	OECD [95b]

[*] Assumptions: Load factor = 75%; operating tails assay = 0.2%; no uranium or plutonium recycling.
[a] PWR = Pressurized Water Reactor; BWR = Boiling Water Reactor; AGR = Advanced Gas-cooled Reactor.
[b] In reference [95b] the parameters are given for a 100 per cent load factor and an operating tails assay of 0.25 per cent. The numbers in the table have been adjusted to correspond to the SIPRI assumptions.

254

Table 8B.3. Demand for enrichment services in the world outside the centrally planned economies (Yugoslavia is included) on a country-by-country basis (in t SWU/yr)*

Country	1980	1985	1990
Austria	81	81	81
Belgium	173	572	572
Brazil		66	327
Canada	29	29	29
Egypt			65
Finland	221	242	242
France	1 697	4 672	5 447
FR Germany	1 024	2 088	2 249
India	47	47	47
Italy	142	595	360
Japan	1 582	3 090	2 648
Korea, South	59	498	777
Mexico		153	153
Netherlands	53	53	53
Philippines		148	65
South Africa		194	194
Spain	67	859	1 408
Sweden	423	1 291	1 060
Switzerland	207	318	426
Taiwan	141	676	554
Turkey			46
UK	243	598	852
USA	5 967	12 502	15 177
Yugoslavia		65	65
Total	**12 156**	**28 837**	**32 897**

* Occasionally the demand for enrichment services is seen to decrease from one time-period to the next (e.g. the Philippines). This is the result of the larger demand in the first operating year of a power reactor than for subsequent years (see table 8B.2).

Table 8B.4. Uranium enrichment capacity available for the world outside the centrally planned economies area (10^3 t SWU/yr)

Supplier	1980	1981	1982	1983	1984	1985	1986	1987	1988	1989	1990	Ref.*
USA, Diffusion	26.4	26.4	27.0	27.3	27.3	27.3	27.3	27.3	27.3	27.3	27.3	[142]
USA, Centrifuge										2.2	2.2	
USSR	3.9	4.1	4.0	4.4	3.4	3.4	3.2	3.3	3.2	2.5	2.4	[143]
Eurodif	6.0	8.4	10.6	10.6	10.6	10.6	10.6	10.6	10.6	10.6	10.6	
Urenco[a]	0.45	0.55	0.7	0.85	1.1	1.4	1.7	2.0	2.0	2.0	2.0	[39b]
Japan[b]	0.03[c]	0.07	0.07	0.07	0.07	0.07 – 0.25	0.07 – 0.25	0.07 – 0.25	0.07 – 0.25	0.07 – 0.25	0.07 – 0.25	[39d]
South Africa[b]	0.006	0.006	0.006	0.006 – 0.3	0.006 – 0.03	0.006 – 0.03	0.006 – 0.03	0.006 – 0.03	0.006 – 0.03	0.006 – 0.03	0.006 – 0.03	
Brazil[b]		0 – 0.006	0 – 0.006	0 – 0.3	0 – 0.3	0 – 0.3	0 – 0.3	0 – 0.3	0 – 0.3	0 – 0.3	0 – 0.3	
France, Chemical exchange						0.05 – 0.1	0.05 – 0.1	0.05 – 0.1	0.05 – 0.1	0.05 – 0.1	0.05 – 0.1	
Total	**36.8**	**39.5**	**42.3**	**43.2 – 43.8**	**42.4 – 43.0**	**42.7 – 43.7**	**42.8 – 43.8**	**43.2 – 44.2**	**43.1 – 44.1**	**44.6 – 45.6**	**44.5 – 45.5**	

* If no reference is given, then see discussion under "Supply of enrichment services" in this appendix, p. 259.
a The numbers are taken from a figure representing Urenco's projected build-up rate and comparing it with the deliveries under signed contracts.
b The limits of the capacities given here differ somewhat from those given in table 8.6 since this table is based on plans which are rather firm.
c The 1980 capacity represents the first two construction phases of the pilot plant. A third phase, with more efficient centrifuges, is scheduled to be completed in 1981.

Table 8B.5. Uranium enrichment production available for the world outside the centrally planned economies area (10^3 t SWU/yr)

Supplier	1980	1981	1982	1983	1984	1985	1986	1987	1988	1989	1990	Ref.*
USA[a]	10.5	9.0	20.5	20.5	21.5	21.5	23.5	26.0	26.5	28.0	29.5	[144]
USSR	3.9	4.1	4.0	4.4	3.4	3.4	3.2	3.3	3.2	2.5	2.4	[143]
Eurodif	6.0	8.4	10.6	10.6	10.6	10.6	10.6	10.6	10.6	10.6	10.6	
Urenco	0.45	0.55	0.7	0.85	1.1	1.4	1.7	2.0	2.0	2.0	2.0	[39b]
Annual production	20.9	22.1	35.8	36.2	36.6	36.9	39.0	41.9	42.3	43.1	44.5	
Cumulative production	**20.9**	**42.9**	**78.7**	**114.9**	**151.5**	**188.4**	**227.4**	**269.3**	**311.6**	**354.7**	**399.2**	

* If no reference is given, then see discussion under "Supply of enrichment services" in this appendix, p. 259.
[a] The numbers are taken from a figure representing the US Department of Energy's Operating Plan.

Table 8B.6. Contracted enrichment services by non-US utilities with the four major suppliers (t SWU)

Recipients		Techsnabexport (USSR)		Eurodif		Urenco		DoE (USA)	
		t SWU	Delivery span	t SWU	Delivery span	t SWU	Delivery span	t SWU	Delivery span
Austria	Committed	1 075	<1979 – 1989	–	–	–	–	398	<1979
Belgium	Committed	1 300	1979 – 1985	8 752	1980 – 1990	–	–	5 002	<1979 – 1995
	Optional	–	–	10 700	1991 – 2000	–	–	–	–
Brazil	Committed	–	–	–	–	3 901	1981 – 1990	1 911	<1979 – 2000
Egypt	Committed	–	–	–	–	–	–	238	1990 – 1995
	Optional	–	–	–	–	–	–	456	1991 – 2000
Finland	Committed	7 441	<1979 – 2000	–	–	–	–	–	–
France	Committed	4 630	<1979 – 1983	47 648	1979 – 1990	–	–	7 006	<1979 – 1988
	Optional	3 000	1984 – 1988	41 100	1991 – 2000	–	–	–	–
FR Germany	Committed	16 547	<1979 – 2000	690	1981 – 1985	12 338	1981 – 2000	18 215	<1979 – 1995
	Optional	5 000	1987 – 2000	–	–	–	–	6 400	1990 – 2000
India	Committed	–	–	–	–	–	–	1 325	<1979 – 1995
Iran	Committed	–	–	8 800	1981 – 1990	–	–	–	–
	Optional	–	–	9 600	1991 – 2000	–	–	–	–
Ireland	Committed	–	–	–	–	831	1991 – 2000	–	–
Italy	Committed	4 225	<1979 – 1983	23 262	1979 – 1990	–	–	2 441	<1979 – 1995
	Optional	–	–	24 000	1991 – 2000	–	–	1 044	1991 – 2000
Japan	Committed	–	–	10 000	1980 – 1989	–	–	68 654	<1979 – 2000
	Optional	–	–	11 000	1990 – 2000	–	–	39 767	1985 – 2000
Korea, South	Committed	–	–	–	–	–	–	4 773	<1979 – 1990
	Optional	–	–	–	–	–	–	9 505	1986 – 2000
Mexico	Committed	–	–	–	–	–	–	1 637	<1979 – 1995
	Optional	–	–	–	–	–	–	1 444	1990 – 2000
Netherlands	Committed	–	–	–	–	–	–	1 931	<1979 – 1995
Philippines	Committed	–	–	–	–	81	<1979 – 1986	435	1982 – 1986
	Optional	–	–	–	–	–	–	1 190	1987 – 2000

South Africa	Committed	–	–	–	–	–	–	2 727	1981 – 1995
	Optional	–	–	–	–	–	–	2 006	1991 – 2000
Spain	Committed	7 484	<1979 – 1990	9 198	1980 – 1990	–	–	7 939	<1979 – 1989
	Optional	–	–	10 700	1991 – 2000	–	–	10 065	1986 – 2000
Sweden	Committed	2 530	1979 – 2000	–	–	–	–	15 426	<1979 – 2000
	Optional	–	–	–	–	–	–	7 837	1987 – 2000
Switzerland	Committed	–	–	729	1982 – 1990	–	–	6 232	<1979 – 1995
	Optional	–	–	–	–	–	–	4 602	1986 – 2000
Taiwan	Committed	–	–	–	–	–	–	6 210	<1979 – 2000
	Optional	–	–	–	–	–	–	6 737	1986 – 2000
Thailand	Committed	–	–	–	–	–	–	162	1990
	Optional	–	–	–	–	–	–	380	1996 – 2000
UK	Committed	1 000	1980 – 1989	–	–	6 816	<1979 – 1995	–	–
Yugoslavia	Committed	–	–	–	–	–	–	667	<1979 – 1986
	Optional	–	–	–	–	–	–	1 320	1986 – 2000
Total committed		**46 232**		**109 079**		**23 967**		**153 329**	
Total optional		**8 000**		**107 100**		–		**92 753**	

< = Prior to.
Source: Reference [59a].

Table 8B.7. Supply and demand of enrichment services in the world outside the centrally planned economies (though Yugoslavia is included) in the 1980s (in 10^3 t SWU)

Year	Demand[a]		Supply[b]		
	Annual	Cumulative	Capacity[c]	Annual production	Cumulative production
1980	12.2	12.2	36.8	20.9	20.9
1981	17.0	29.2	39.5	22.1	42.9
1982	19.4	48.6	42.3	35.8	78.7
1983	23.0	71.6	43.2	36.2	114.9
1984	26.4	98.0	42.4	36.6	151.5
1985	28.8	126.8	42.7	36.9	188.4
1986	29.1	155.9	42.8	39.0	227.4
1987	30.2	186.1	43.2	41.9	269.3
1988	32.7	218.9	43.1	42.3	311.6
1989	33.2	252.1	44.6	43.1	354.7
1990	32.8	284.9	44.5	44.5	399.2

[a] See discussion under 'Demand for enrichment services' in this appendix.
[b] See discussion under 'Supply of enrichment services' in this appendix.
[c] The lower limit of table 8B.3 is used.

References to Part One

1. Weast, R.C. (ed.), *Handbook of Chemistry and Physics*, 58th ed., 1977–78 (CRC Press, Cleveland, Ohio, 1977), Table of the isotopes, pp. B-271–B-355.

2. Cowan, G.A., 'A natural fission reactor', *Scientific American*, Vol. 235, No. 1, July 1976, pp. 36–47.

3. *Nuclear Proliferation Factbook*, US Senate Committee on Governmental Affairs (US Government Printing Office, Washington, D.C., September 1980).
   - (a) —, p. 175.
   - (b) —, pp. 407–408.

4. Willrich, M. and Taylor, T.B., *Nuclear Theft: Risks and Safeguards* (Ballinger, Cambridge, Mass., 1974).
   - (a) —, p. 16.
   - (b) —, p. 17.
   - (c) —, p. 125.
   - (d) —, p. 127.
   - (e) —, p. 78.
   - (f) —, p. 128.

5. Inglis, D.R., *Nuclear Energy: Its Physics and Its Social Challenge* (Addison–Wesley, Reading, Mass., 1973), pp. 314–15.

6. *The Structure and Content of Agreements between the Agency and States Required in Connection with the Treaty on the Non-Proliferation of Nuclear Weapons*, INFCIRC/153 (IAEA, Vienna, 1971).
   - (a) —, para. 11.
   - (b) —, para. 28.
   - (c) —, para. 29.
   - (d) —, para. 31.
   - (e) —, para. 70–84.
   - (f) —, para. 42–48.

7. Allison, G.T., *Essence of Decision* (Little Brown, Boston, Mass., 1971).

8. Hewlett, R.G. and Anderson, Jr., O.E., *The New World: 1939/1946* (Pennsylvania State University Press, University Park, Pa., 1962).
   - (a) —, p. 646.
   - (b) —, p. 168.
   - (c) —, p. 301.
   - (d) —, p. 624.
   - (e) —, pp. 130–31.
   - (f) —, p. 141.

9. Kramish, A., *Atomic Energy in the Soviet Union* (Stanford University Press, Stanford, Calif., 1959), pp. 110, 115.

10. Wilkie, T., 'Tricastin points the road to energy independence', *Nuclear Engineering International*, Vol. 25, No. 305, October, 1980, p. 41.

11. London, H., *Separation of Isotopes* (Newnes, London, 1961).
   - (a) —, p. 249.
   - (b) —, p. 433.
   - (c) —, p. 474.

12. Irving, D., *The German Atomic Bomb* (Simon & Schuster, New York, 1967), pp. 127–28.

13. *International Nuclear Fuel Cycle Evaluation (INFCE)* (IAEA, Vienna, 1980).

References to Part One (cont.)

(a) —, INFCE/PC/2/2, Enrichment Availability, Chapter 7.
(b) —, ibid., pp. 146–47.
(c) —, INFCE/PC/2/9, Summary volume, p. 33.
(d) —, ibid., p. 51.
(e) —, INFCE/PC/2/2, Enrichment Availability, pp. 125–26.
(f) —, ibid., p. 146.
(g) —, INFCE/PC/2/9, Summary Volume, p. 45.
(h) —, INFCE/PC/2/2, Enrichment Availability, p. 141.
(i) —, INFCE/PC/2/9, Summary Volume, p. 1.
(j) —, ibid., p. 38

14. *Nuclear Proliferation and Civilian Nuclear Power, Report of the Non-Proliferation Alternative Systems Assessment Program*, Vol. II, DoE/NE-0001/2, Section 3.1 (US Department of Energy, Washington, D.C., June 1980).

15. US Congress, Office of Technology Assessment, *Nuclear Proliferation and Safeguards* (Praeger, New York, 1977).
(a) —, p. 209.
(b) —, p. 219.
(c) —, pp. 209–11.

16. Zippe, G., 'The development of short-bowl ultracentrifuges', *University of Virginia Report* No. EP-4420–101–60U, submitted to Physics Branch, Division of Research, US Atomic Energy Commission, Washington, D.C., July 1960.

17. Smith, D., *South Africa's Nuclear Capability* (World Campaign Against Military and Nuclear Collaboration with South Africa and UN Centre Against Apartheid, London, February 1980).

18. Grant, W.L., Wannenburg, J.J. and Haarhoff, P.C., 'Cascade technique for the South African enrichment process', *AIChE Symposium Series*, Vol. 73, No. 169, 1977.

19. Gowing, M., *Britain and Atomic Energy 1939–1945* (St. Martins, New York, 1964).

20. Pierce, A.J., *Nuclear Politics* (Oxford University Press, Oxford, 1972).

21. Weart, S., *Scientists in Power* (Harvard University Press, Cambridge, Mass., 1979).

22. Goldschmidt, B., *Les Rivalités Atomiques* (Fayard, Paris, 1967).

23. Goldschmidt, B., *Le Complexe Atomique* (Fayard, Paris, 1980).
(a) —, pp. 304–20.
(b) —, pp. 387–88.
(c) —, pp. 310–12, 402, 442–43.

24. Kohl, W.L., *French Nuclear Diplomacy* (Princeton University Press, Princeton, N.J., 1971).

25. Nieburg, H.L., *Nuclear Secrecy and Foreign Policy* (Public Affairs Press, Washington, D.C., 1964).

26. 'Forecast of growth of nuclear power', WASH 1139 (USAEC, Washington, D.C., January 1971).

27. 'Uranium: resources, production and demand', (OECD-NEA/IAEA, Paris, 1970).

28. *Agreement between the Kingdom of the Netherlands, the Federal Republic of Germany and the United Kingdom of Great Britain and Northern Ireland on Collaboration in the Development and Exploitation of the Gas Centrifuge Process for Producing Enriched Uranium* (Almelo Treaty), Tractatenbladen van het

References to Part One (cont.)

Koninkrijk der Nederlanden, Jaargang 1970, no. 41 (Staatsuitgeverij, The Hague, 1970).

    (a) —, Art. III.

    (b) —, Art. VI.

    29. Willrich, M., *Non-Proliferation Treaty: Framework for Nuclear Arms Control* (The Michie Corp., Charlottesville, Va., 1965), p. 165.

    30. SIPRI, *Safeguards Against Nuclear Proliferation* (MIT Press, Cambridge, Mass., 1975).

    (a) —, p. 2.

    (b) —, pp. 32–42.

    (c) —, p. 40.

    (d) —, pp. 48–49.

    31. 'A report on the international control of atomic energy', (Washington, D.C., 16 March 1946), in *Reader on Nuclear Non-Proliferation*, prepared for the Subcommittee on Energy, Nuclear Proliferation and Federal Services, of the Committee on Governmental Affairs, US Senate, by the Congressional Research Service, Library of Congress, Washington, D.C., December 1980.

    (a) —, p. 19.

    (b) —, p. 17.

    32. *Nuclear Weapons Proliferation and the IAEA, An Analytical Report*, prepared for the Committee on Governmental Operations, US Senate, by the Congressional Research Service (US Government Printing Office, Washington, D.C., March 1976).

    33. *Safeguarding Nuclear Materials, Proceedings of a Symposium*, Vienna, 20–24 October 1975 (IAEA, Vienna, 1976).

    (a) —, Dragnev, T.N., de Carolis, M., Keddar, A., Konnov, Yu., Martinez-Garcia, G. and Waligura, A.J., 'Some Agency contributions to the development of instrumental techniques in safeguards', Vol. II, pp. 37–43.

    34. Information Circular INFCIRC/66/Rev 2 (IAEA, Vienna, 1968), Annex II, para. 12.

    35. *The Text of the Safeguards Agreement of 26 February 1976 between the Agency, Brazil, and the Federal Republic of Germany*, INFCIRC/237 (IAEA, Vienna, 26 May 1976).

    36. *IAEA Safeguards Glossary* (IAEA, Vienna, 1980).

    (a) —, p. 21, para. 89.

    (b) —, p. 20, para. 87.

    (c) —, p. 22, para. 90.

    (d) —, p. 21, table II.

    (e) —, pp. 28–29, para. 110.

    (f) —, p. 33, para. 126.

    (g) —, p. 30, para. 113.

    (h) —, p. 25, para. 100, table IV.

    (i) —, p. 50, para. 216.

    (j) —, p. 50, para. 217.

    37. Younkin, J.M., *Projected Uranium Measurement Uncertainties for the Gas Centrifuge Enrichment Plant* (Union Carbide Corp., Nuclear Division, Oak Ridge Y-12 Plant, Oak Ridge, Tenn., Y/DS-100, February 1979).

    (a) —, p. 4.

    (b) —, p. 7.

References to Part One (cont.)

(c) —, p. 10.

38. Blumkin, S., and von Halle, E., *The Behaviour of the Uranium Isotopes in Separation Cascades* (Union Carbide Corp., Nuclear Division, Oak Ridge Gaseous Diffusion Plant, Oak Ridge, Tenn., K-1839, Part 1, 21 November 1972; Part 2, 29 August 1973; Part 3, 22 March 1974; Part 4, 18 December 1974; Part 5, 19 January 1976).

(a) —, Part 5, pp. 30–33.

(b) —, Part 5, p. 28.

39. Younkin, J.M. and Rushton, J.E., *Effect of Short-term Material Balances on the Projected Uranium Measurement Uncertainties for the Gas Centrifuge Enrichment Plant* (Union Carbide Corp., Nuclear Division, Oak Ridge, Tenn., Y/DS-109, December 1979), p. 6.

40. Kratzer, M.B., 'Prospective trends in international safeguards', *The Atlantic Papers, Nuclear Non-Proliferation and Safeguards*, No. 42 (Atlantic Institute for International Affairs, Paris, April 1981), p. 38.

41. Bahm, W., Gupta, D., Didier, H.J. and Wappner, J., 'Nuclear materials management in a uranium enrichment facility based on the separation nozzle process', *Nuclear Materials Management*, Vol. 7, 27–29 June 1978, pp. 337–47.

(a) —, p. 341.

42. Blumkin, S. and von Halle, E., *A Method for Estimating the Inventory of an Isotope Separation Cascade by the Use of Minor Isotope Transient Concentration Data* (Union Carbide Corp., Nuclear Division, Oak Ridge Gaseous Diffusion Plant, Oak Ridge, Tenn., K-1892, 1 January 1978).

(a) —, p. 27.

(b) —, p. 12.

43. Weisz, G., Director, Office of Safeguards and Security Defense Programs, US Department of Energy, Washington D.C., private communication, October 1981.

44. SIPRI, *Nuclear Energy and Nuclear Weapon Proliferation* (Taylor & Francis, London, 1979).

(a) —, p. 213.

(b) —, pp. 283–84.

(c) —, p. 282.

45. 'International symposium on the safeguarding of nuclear material', *Nuclear Engineering International*, Vol. 20, No. 237, December 1975, p. 1022.

46. Kissinger, H., *White House Years* (Little, Brown, Boston, Mass., 1979), p. 554.

47. Russell, B., quoted in *New York Times*, 21 November 1948, p. 4.

48. 'Delhi attack on atomic facility feared: Pakistan prepares to defend its bomb', *Financial Times*, 14 August 1979.

49. 'Atombomben für den Islam', *Der Spiegel*, 12 November 1979, pp. 202–209.

50. SIPRI, *World Armaments and Disarmament, SIPRI Yearbook 1977* (Almqvist & Wiksell, Stockholm, 1977).

(a) —, p. 20.

(b) —, p. 22.

51. SIPRI, *Internationalization to Prevent the Spread of Nuclear Weapons* (Taylor & Francis, London, 1980).

(a) —, p. 111.

References to Part One (cont.)

(b) —, p. 19.
(c) —, p. 191.
(d) —, p. 185.
(e) —, p. 14.
(f) —, p. 190.
(g) —, p. 18.
(h) —, p. 193.
(i) —, p. 194.
(j) —, p. 199.
(k) —, p. 5.
(m) —, p. 8.
(n) —, p. 73.
(o) —, p. 200.
(p) —, p. 9.

52. Nuclear Suppliers Group, *Guidelines for Nuclear Transfer*, 1978 (available from the Department of Foreign Affairs, The Hague).

(a) —, Guidelines, including annexes A and B.
(b) —, Part B.

53. 'Carter to overrule US Nuclear Fuel Panel, send 38 tons of atomic fuel to India', *International Herald Tribune*, 20 June 1980, p. 1.

54. 'Carter lobbies Senators for Uranium sale to India', *International Herald Tribune*, 25 September 1980.

55. 'US cautions India on atomic safeguards', *International Herald Tribune*, 26 September 1980, p. 2.

56. *Nuclear Proliferation: The Situation in Pakistan and India*, Hearing before the Subcommittee on Energy, Nuclear Proliferation, and Federal Services, Committee on Governmental Affairs, US Senate, Washington, D.C., 1 May 1979.

57. Casper, B.M., *Bulletin of the Atomic Scientists*, Vol. 33, No. 1, January 1977, pp. 28–41.

58. Chayes, A. and Lewis, W. (eds.), *International Arrangements for Nuclear Fuel Reprocessing* (Ballinger, Cambridge, Mass., 1977).

(a) —, p. 73.
(b) —, p. 74.

59. *Regional Nuclear Fuel Cycle Centers*, Vol. I (IAEA, Vienna, 1977).

(a) —, p. 59–60.
(b) —, p. 59.
(c) —, p. 22.
(d) —, p. 63.
(e) —, p. 62.

60. Wonder, E.F., *Nuclear Fuel and American Foreign Policy: Multilateralization for Uranium Enrichment, Atlantic Council Policy Paperback* (Westview Press, Boulder, Colorado, 1977).

(a) —, p. 15.
(b) —, p. 20.

61. Mendershausen, H., *International Cooperation in Nuclear Fuel Services: European and American Approaches* (Rand Corp., Santa Monica, Cal. RAND/P-6308, 1978).

(a) —, p. vi.
(b) —, p. 25.

References to Part One (cont.)

(c) —, p. 34.

(d) —, pp. 73–74.

(e) —, pp.74–75.

62. Smart, I., *Multinational Arrangements for the Nuclear Fuel Cycle*, Department of Energy, Energy Paper No. 43 (H.M.S.O., London, 1980).

(a) —, chapter VI.

(b) —, p. 54.

(c) —, p. 14.

(d) —, p. 40.

(e) —, p. 38.

(f) —, chapter IV.

63. *Nuclear Power: Issues and Choices* (MITRE Corp., Ballinger, Cambridge, Mass., 1977), p. 297.

64. Letter from Minister of Economic Affairs to Dutch Parliament, 2 February 1979, Kamerstuk 14261 UltraCentrifuge Project, No. 43.

65. *Eurodif Treaty* (Eurodif S.A., Bagneux, France).

(a) —, Art. XIII.

(b) —, Art. XIV.

66. Petit, J.F., 'Enrichissement de l'uranium par le procédé de diffusion gazeuse', IAEA-CN-36/552, *Nuclear Power and Its Fuel Cycle*, Proceedings of an International Conference, Salzburg, 2–13 May 1977, Vol. 3 (IAEA, Vienna, 1977), p. 126.

67. 'La Société Technicatome retient en France 80 tonnes d'uranium Iranien', *Le Monde*, 23 April 1981, p. 40.

68. *Jaarboek van Buitenlandse Zaken 1973–74*, p. 27B, (Staatsuitgeverij, The Hague, 1974).

69. *Euratom Treaty*, Rome, 25 March 1957, (Verdrag ter oprichting van de Europese Gemeenschap voor Atoomenergie (Euratom), N.V. Uitgeversmij W.C.J. Tjeenk Willink, Zwolle, 1981)

(a) —, Art. 52, 57.

(b) —, Art. 52, 64.

(c) —, ch. VII, VIII.

70. Nau, H.R., *National Politics and International Technology* (Johns Hopkins University Press, Baltimore, Md., 1974).

(a) —, p. 100.

(b) —, p. 100–101.

71. SIPRI, *The NPT: The Main Political Barrier to Nuclear Proliferation* (Taylor & Francis, London, 1980), p. 31.

72. Baruch, B.M., 'United States Atomic Energy Proposals', Address before the United Nations Atomic Energy Commission, 14 June 1946. *Reader on Nuclear Non-Proliferation* (Congressional Research Service, Library of Congress, 7 December 1978).

(a) —, pp. 4, 7.

(b) —, p. 4.

(c) —, p. 7.

(d) —, p. 8.

(e) —, p. 9.

73. Letter of D.E. Lilienthal to G.T. Seaborg, Chairman of USAEC, 2 December 1964, in *Export Reorganization Act of 1976*, Hearings before the

References to Part One (cont.)

Committee on Governmental Operations, US Senate, Washington, D.C., 19, 20, 29, 30 January and 9 March, 1976, p. 12.

74. Epstein, W., *The Last Chance* (The Free Press, New York, 1976).

(a) —, chapters V and VII.

(b) —, p. 104.

(c) —, p. 237.

75. 'How Israel got the bomb', *Time Magazine*, Special Report, 12 April 1976, p. 39.

76. SIPRI, *World Armaments and Disarmament, SIPRI Yearbook 1978* (Taylor & Francis, London, 1978), p. 73.

(a) —, p. 36.

(b) —, p. 73.

77. 'Neutron bomb is suspected in South African explosion', *International Herald Tribune*, 15/16 March 1980, p. 6.

78. Torrey, L., 'Is South Africa a nuclear power?', *New Scientist*, 24 July 1980.

79. SIPRI, *World Armaments and Disarmament, SIPRI Yearbook 1980* (Taylor & Francis, London, 1980), pp. 198, 360.

80. *Nuclear Non-Proliferation Act of 1978*, Public Law 95–242, 10 March 1978, section 104.

81. Galois, P., *The Balance of Terror: Strategy for the Nuclear Age* (Riverside Press, New York, 1961).

82. Waltz, K.N., 'The spread of nuclear weapons: more may be better', *Adelphi Paper* No. 171 (International Institute for Strategic Studies, London, 1981).

83. Lovins, A.B. and Lovins, L.H., *Energy/War: Breaking the Nuclear Link* (Harper & Row, New York, 1980).

(a) —, chapter 2.

(b) —, chapters 4, 5.

84. SIPRI, *The Law of War and Dubious Weapons* (Almqvist & Wiksell, Stockholm, 1976), p. 50.

85. 'US rules led Australia to reject atomic deal', *International Herald Tribune*, 8 October 1982, p. 4.

References to Part Two

1. Weast, R.C. (ed.), *Handbook of Chemistry and Physics*, 58th ed., 1977–78 (CRC Press Inc., Cleveland, Ohio, 1977).

  (a) —, p. B-56.

  (b) —, p. B-55.

  (c) —, p. B-142.

2. Villani, S. (ed.), *Topics in Applied Physics*, Vol. 35, *Uranium Enrichment* (Springer-Verlag, New York, 1979).

  (a) —, B. Brigoli, 'Cascade theory', p. 15.

  (b) —, B. Brigoli, p. 41.

  (c) —, B. Brigoli, p. 15.

  (d) —, B. Brigoli, p. 14.

  (e) —, B. Brigoli, p. 16.

  (f) —, B. Brigoli, p. 25.

  (g) —, B. Brigoli, p. 44.

  (h) —, D. Massignon, 'Gaseous diffusion', pp. 156–58.

  (i) —, B. Brigoli, p. 22.

  (j) —, D. Massignon, p. 123.

  (k) —, D. Massignon, p. 125.

  (m) —, D. Massignon, p. 126.

  (n) —, D. Massignon, p. 127.

  (o) —, D. Massignon, p. 92.

  (p) —, D. Massignon, pp. 167, 170.

  (q) —, D. Massignon, p. 133.

  (r) —, D. Massignon, p. 117.

  (s) —, Soubbaramayer, 'Centrifugation', p. 183.

  (t) —, Soubbaramayer, p. 235.

  (u) —, S. Villani, 'Review of separation processes', p. 7.

  (v) —, D. Massignon, equation 3.181, p. 124.

  (w) —, E.W. Becker, 'Separation nozzle', p. 254.

  (x) —, E.W. Becker, p. 257.

  (y) —, B. Brigoli, pp. 31–39.

  (z) —, B. Brigoli, pp. 37–39.

  (aa) —, E.W. Becker, p. 260.

  (bb) —, C.P. Robinson and R.J. Jensen, 'Laser methods of uranium isotope separation', p. 274.

  (cc) —, C.P. Robinson and R.J. Jensen, p. 283.

  (dd) —, C.P. Robinson and R.J. Jensen, p. 277.

  (ee) —, C.P. Robinson and R.J. Jensen, p. 288.

  (ff) —, F. Boeschoten and N. Nathrath, 'Plasma separating effects', p. 311.

  (gg) —, F. Boeschoten and N. Nathrath, pp. 310–11.

  (hh) —, F. Boeschoten and N. Nathrath, p. 294.

  (ii) —, F. Boeschoten and N. Nathrath, p. 291.

  (jj) —, F. Boeschoten and N. Nathrath, p. 300.

  (kk) —, F. Boeschoten and N. Nathrath, p. 303.

  (mm) —, F. Boeschoten and N. Nathrath, p. 299.

3. Cohen, K., *The Theory of Isotope Separation as Applied to the Large Scale Production of U^{235}* (McGraw-Hill, New York, 1951).

  (a) —, p. xv.

References to Part Two (cont.)

(b) —, p. 26.
(c) —, p. 27.
(d) —, pp. 31–34.

4. SIPRI, *Internationalization to Prevent the Spread of Nuclear Weapons* (Taylor & Francis, London, 1980), J. Rotblat, 'Background data relating to the management of nuclear fuel cycle materials and plants', p. 35.

5. Rose, A. and Rose, E., *Techniques of Organic Chemistry*, Vol. IV, 'Distillation' (Interscience Publishing Inc., New York, 1951), pp. 2–9.

6. Janes, G.S. *et al.*, 'Two-photon laser isotope separation of atomic uranium: spectroscopic studies, excited-state lifetimes, and photoionization cross sections' *IEEE Journal of Quantum Electronics*, Vol. QE-12, 1976, pp. 111–20.

7. *Van Nostrand's Scientific Encyclopedia*, 5th ed. (Van Nostrand Reinhold Co., New York, 1976), p. 2263.

8. London, H., *Separation of Isotopes* (Newnes, London, 1961), p. 447.

9. Louck, J.K. and Galbraith, H.W., 'Eckart vectors, Eckart frames, and polyatomic molecules' *Reviews of Modern Physics*, Vol. 48, 1976, p. 69.

10. Roux, A.J.A. *et al.*, 'Development and progress of the South African enrichment project', *Nuclear Power and Its Fuel Cycle*, Proceedings of an International Conference, Salzburg, 2–13 May 1977 (IAEA, Vienna, 1977), p. 182.

11. London, H., *Separation of Isotopes* (Newnes, London, 1961).

(a) —, p. 332.
(b) —, p. 112.
(c) —, pp. 12–13.
(d) —, p. 452.
(e) —, p. 449.
(f) —, p. 447.

12. Wilkie, T., 'Tricastin points the road to energy independence', *Nuclear Engineering International*, Vol. 25, No. 305, October 1980, p. 44.

13. *Nuclear Power and Its Fuel Cycle*, Proceedings of an International Conference, Salzburg, 2–13 May 1977, Vol. 3 (IAEA, Vienna, 1977).

(a) —, T. Aochi, R. Nakane and S. Takahashi, 'Research on and economic evaluation of uranium enrichment by gaseous diffusion in Japan', pp. 127–41.

(b) —, T. Aochi, R. Nakane and S. Takahashi, p. 130, figure 1.

(c) —, W.L. Grant, from a discussion following presentation of paper by A.J.A. Roux, W.L. Grant, R.A. Barbour, R.S. Loubsec, and J.J. Wannenburg entitled 'Development and progress of the South African enrichment project', p. 182.

(d) —, A.J.A. Roux *et al.*, 'Development and progress of the South African enrichment project', p. 178.

(e) —, A.J.A. Roux *et al.*, p. 179.

(f) —, A.J.A. Roux *et al.*, p. 182.

(g) —, C. Frejacques *et al.*, 'Evolution des procédés de séparation des isotopes de l'uranium en France', pp. 209–11.

14. Kabayama, S. and Ikeda, H., 'Process for producing a porous diffusion membrane', Japanese patent document 1975–4627/B, Abstract 260517, *INIS Atom Index 4* (IAEA, Vienna, 1976), p. 3807.

15. *Proceedings of International Conference on Uranium Isotope Separation*, British Nuclear Energy Society in conjunction with Kerntechnische Gesellschaft, London, 5–7 March 1975.

References to Part Two (cont.)

(a) —, C.C. Hopkins, 'Operating experience with US gaseous diffusion plants', Paper No. 14, p. 3.

(b) —, H. Geppert, 'The industrial implication of the separation nozzle process', Paper No. 3, table 3, p. 5.

(c) —, C.J. Hardy, 'Recent experimental and assessment studies of uranium enrichment by ion exchange', Paper No. 12, p. 4.

(d) —, J. Chatelet *et al.*, 'Chemical exchange between UF_6 and UF_6 ion in anhydrous hydrofluoric acid', Paper No. 30, p. 2.

(e) —, J. Chatelet *et al.*, p. 5.

(f) —, Papers No. 7–11: M. Hashmi and A.J. van der Houven van Oordt, 'One-step separation of isotopes using a uranium plasma', Paper No. 7.
N. Nathrath, H. Kress, J. McClure, G. Mück, M. Simon and H.J. Dibbert, 'Isotope separation in rotating plasmas', Paper No. 8.
J.B.S. Cairns, 'Isotopic separation in a rotating neon plasma', Paper No. 9.
B.W. James, D.D. Millar and S.W. Simpson, 'The separation of neon isotopes by a plasma centrifuge', Paper No. 10.
F. Boeschoten, 'On the possibility to separate isotopes by means of a cylindrical plasma column rotating around its axis', Paper No. 11.

(g) —, B.W. James, D.D. Millar and S.W. Simpson, 'The separation of neon isotopes by a plasma centrifuge', Paper No. 10, p. 2.

16. Brown, A.C. and MacDonald, C.B. (eds.), *The Secret History of the Atomic Bomb* (Dial Press/James Wade, New York, 1977), p. 312.

17. 'Enrichment plant compressors for 20-years' non-stop operation', *Nuclear Engineering International*, Vol. 24, No. 290, September 1979.

(a) —, p. 33.

(b) —, p. 34.

18. Villani, S. *Isotope Separation* (American Nuclear Society, Chicago, Illinois, 1976).

(a) —, p. 176.

(b) —, pp. 49–61.

(c) —, p. 61.

19. Semler, E.G. (ed.), *Advancing Technologies* (Mechanical Engineering Publications Ltd., London, 1977).

(a) —, D. Aston, 'The ultra-high-speed centrifuge', pp. 60–70.

(b) —, D. Aston, p. 64.

20. Avery, D.G. and Davies, E., *Uranium Enrichment by Gas Centrifuge* (Mills and Boon, London, 1973).

(a) —, p. 25.

(b) —, p. 44.

21. Olander, D.R., 'The gas centrifuge', *Scientific American*, Vol. 239, No. 2, August 1978, pp. 27–33.

22. Kanagawa, A., 'Evaluation of enrichment by centrifugal separation', *Genshiryoku Kogyo*, Vol. 20, No. 2, 1974, pp. 40–45 [Engl. transl. UCRL-Transl-10891, July 1975, p. 10].

23. *International Nuclear Fuel Cycle Evaluation, INFCE/PC/2/2, Enrichment Availability* (IAEA, Vienna, 1980).

(a) —, p. 132.

(b) —, p. 133.

(c) —, p. 45.

24. Henley, E.J. and Lewins, J. (eds.), *Advances in Nuclear Science and Technology*, Vol. 6 (Academic Press, New York, 1972).

(a) —, D.R. Olander, 'Technical basis of the gas centrifuge', pp. 136–39.

(b) —, D.R. Olander, p. 138.

25. Crossley, P.S., 'Centrifuges help match supply and demand', *Nuclear Engineering International*, Vol. 25, No. 305, October 1980, p. 47.

26. Eaton, R.R., Fox, R.L. and Touryan, K.J., 'Isotope enrichment by aerodynamic means: a review and some theoretical considerations', *Journal of Energy*, Vol. 1, No. 4, July/August 1977.

(a) —, p. 229.

(b) —, pp. 230–31.

27. Benedict, M. *et al.*, *Report of Uranium Isotope Separation Review Ad Hoc Committee*, Oak Ridge National Laboratory Report, ORO-694 (Oak Ridge, Tenn., 2 June 1972).

(a) —, chapter 7.

(b) —, section 7.2.

(c) —, chapter 4.

(d) —, section 4.2.

28. *Nuclear Energy Maturity* (Pergamon Press, Oxford, 1975).

(a) —, Compargue, R. *et al.*, 'On aerodynamic separation methods', pp. 5–25.

(b) —, A.J.A. Roux, 'The South African Enrichment Project', p. 40.

29. Berkhahn, W., Ehrfeld, W. and Krieg, G., 'Influence of flowfield structure on uranium isotope separation in the separation nozzle', *Nuclear Technology*, Vol. 40, No. 3, October 1978, pp. 329–40.

30. Becker, E.W. *et al.*, 'Present state and development potential of separation nozzle process', revised version of a report presented by W. Ehrfeld at the Gordon Research Conference on Physics and Chemistry of Isotopes, Asilomar, Pacific Grove, California, 1–5 July 1974 (Gesellschaft für Kernforschung mbH, Karlsruhe, FRG, 1974), KFK-2067, p. 2.

31. Ehrfeld, W., private communication, December 1981.

32. Becker, E.W., Nogueira-Batista, P. and Völcker, H., 'Uranium enrichment by the separation nozzle method within the framework of German/Brazilian cooperation', *Nuclear Technology*, Vol. 52, No. 1, January 1981, pp. 105–14.

33. *Nuclear Proliferation and Civilian Nuclear Power*, Report of the Nonproliferation Alternative Systems Assessment Program, Vol. II, US Department of Energy, DoE/NE-0001/2, June 1980, pp. 3–12.

34. Grant, W.L., Wannenburg, J.J. and Haarhoff, P.C., 'The cascade technique for the South African enrichment process', *AIChE Symposium Series*, Vol. 73, No. 169, 1977.

(a) —, p. 24.

(b) —, p. 21.

(c) —, p. 23.

35. Ashford, N., 'South Africans now able to produce A-bomb', *London Times*, 30 April 1981, p. 1.

36. Hutchison, C.A., Jr., in *Chemical Separation of the Uranium Isotopes*, G.M. Murphy (ed.), US Atomic Energy Commission, TID-5224 (Oak Ridge, Tenn., 1952).

37. Coates, J.H. *et al.*, 'Chemical exchange: commercial scale in ten years?', *Nuclear Engineering International*, Vol. 25, No. 305, October 1980, p. 49.

References to Part Two (cont.)

38. Seko, M. and Miyake, T., 'Chemical enrichment process in Japan', *Atoms in Japan*, July 1980.
    (a) —, pp. 20–23.
    (b) —, p. 20.
    (c) —, p. 21.
    (d) —, p. 23.
39. Fujii, Y., Fukuda, J. and Kakihana, H., 'Separation of uranium isotopes using ion-exchange chromatography', *Journal of Nuclear Science and Technology* (Tokyo), Vol. 15, No. 10, October 1978, pp. 745–52.
40. Seko, M. *et al.*, 'Uranium isotope enrichment by chemical method', *Nuclear Technology,* Vol. 50, No. 2, September 1980.
    (a) —, pp. 178–86.
    (b) —, p. 180.
    (c) —, p. 182.
    (d) —, p. 184.
    (e) —, p. 181.
    (f) —, p. 183.
    (g) —, p. 186.
41. *The Kirk-Othmer Encyclopedia of Chemical Technology* (Interscience, London, 1965).
    (a) —, 2nd ed., Vol. 8, p. 747.
    (b) —, 3rd ed., Vol. 5, pp. 339–67.
    (c) —, 3rd ed., Vol. 5, pp. 363–65.
42. *Laser Isotope Separation: Proliferation Risks and Benefits,* Report of the Laser Enrichment Review Panel to JNAI, (Jersey Nuclear-Avco Isotopes), Vol. 2, 27 February 1979.
    (a) —, Appendix G, p. G-20.
    (b) —, Appendix G, p. G-19.
    (c) —, Appendix D, p. D-9.
    (d) —, Appendix D, p. D-6.
    (e) —, Appendix D, pp. D-4–D-5.
    (f) —, P.L. Auer, Appendix F-1.
43. Pedersen, C.J., 'Cyclic polyethers and their complexes with metal salts', *Journal of the American Chemical Society,* Vol. 89, No. 26, 20 December 1967, p. 7018.
44. Betts, R.H. and Bron, J., 'A discussion of partial isotope separation by means of solvent extraction', *Separation Science,* Vol. 12, No. 6, 1977, pp. 635–39.
45. Frejacques, C., Lerat, J.M. and Plurien, P., 'French chemical exchange process', *Symposium on Separation Science and Technology for Energy Applications,* Gatlinburg, Tenn., 30 October–2 November 1979, CEA-Conf. 5062.
    (a) —, figure 2.
    (b) —, figure 3.
    (c) —, p. 4.
46. Coates, J.H., 'Le procédé chimique français d'enrichissement de l'uranium', *CEA Notes d'Information,* October 1980, No. 10, pp. 3–10.
    (a) —, p. 4.
    (b) —, p. 8.
    (c) —, p. 7.
47. Saraceno, A.J. and Trivisonno, C.F. (eds.), *Uranium Isotope Separation*

by Chemical Exchange Reactions between UF₆ and UF–Nitrogen Oxide Complexes, GAT-674 (Goodyear Atomic Corp., Piketon, Ohio, 4 February 1972).

48. Coates, J.H., 'A short briefing on Chemex' (Division d'Etudes de Separation Isotopique et Chimie Physique, Commissariat à l'Energie Atomique [CEA], Paris, unpublished paper, February 1981).

 (a) —, p. 6.
 (b) —, p. 3.
 (c) —, p. 2.
 (d) —, p. 4.

49. *Report of the Energy Research Advisory Board Study Group on Advanced Isotope Separation* (US Department of Energy, Washington, D.C., November 1980).

 (a) —, p. 8.
 (b) —, p. 30.
 (c) —, p. 3, para. 1.2(a).
 (d) —, p. 21.
 (e) —, p. 31.
 (f) —, p. 7.
 (g) —, p. 32.

50. SIPRI, *Nuclear Energy and Nuclear Weapon Proliferation* (Taylor & Francis, London, 1979).

 (a) —, J.H. Coates and B. Barré, 'Practical suggestions for the improvement of proliferation resistance within the enriched uranium fuel cycle', pp. 49–53.

 (b) —, B.T. Feld, 'Can plutonium be made weapon-proof?', pp. 113–19.

51. Forsen, H.K., 'The Exxon/Avco program in laser enrichment: technology status, economic potential, and future prospects', Paper presented at the International Conference on the Nuclear Fuel Cycle, Amsterdam, 14–17 September 1980 (preprint).

 (a) —, figure 1.
 (b) —, figure 3.
 (c) —, figure 2.
 (d) —, p. 7.
 (e) —, figure 7, p. 8.

52. Böhm, H-D.V., Michaelis, W. and Weitkamp, C., 'Hyperfine structure and isotope shift measurements on ^{235}U and laser separation of uranium isotopes by two-step photoionization', *Optics Communications,* Vol. 26, No. 2, August 1978, p. 178.

53. Miller, S.S. *et al.,* 'Infrared photochemistry of a volatile uranium compound with 10μ absorption', *Journal of the American Chemical Society,* Vol. 101, No. 4, 14 February 1979, pp. 1036–37.

54. Cox, D.M. *et al.,* 'Isotope selectivity of infrared laser-driven unimolecular dissociation of a volatile uranyl compound', *Science,* Vol. 205, No. 4404, 27 July 1979, pp. 390–94.

55. 'Jersey Nuclear quits laser enrichment because DoE offers too little support', *Laser Focus,* Vol. 17, No. 5, May 1981, pp. 42–44.

56. Davis, J.I., 'Atomic Vapor Laser Isotope Separation at Lawrence Livermore National Laboratory', UCRL-83516, Conference on Laser and Electro-Optical Systems, San Diego, Calif., 26–28 February 1980.

 (a) —, p. 23.

(b) —, p. 13.

57. Krass, A.S., 'Laser enrichment of uranium: the proliferation connection', *Science,* Vol. 196, 13 May 1977.

(a) —, p. 723.

(b) —, p. 724.

(c) —, table 1, p. 724.

58. Jensen, R.J. *et al.,* 'Prospects for uranium enrichment', *Laser Focus,* Vol. 12, No. 5, table 1, p. 51.

59. Moore, C.B. (ed.), *Chemical and Biochemical Applications of Lasers,* Vol. III (Academic Press, New York, 1977).

(a) —, V.S. Letokhov and C.B. Moore, 'Laser isotope separation', pp. 28–31.

(b) —, R.V. Ambartzumian and V.S. Letokhov, 'Multiple photon infrared laser photochemistry', pp. 167–316.

60. Garwin, R.L., 'The promise of laser isotope separation', *Bulletin of the Atomic Scientists,* Vol. 33, No. 8, October 1977, pp. 8–9.

61. *Nuclear Proliferation and Safeguards,* US Congress, Office of Technology Assessment (Praeger, New York, 1977), p. 183.

62. Levin, L.A., Janes, G.S. and Levy, R.H., 'Suppression of unwanted lasing in laser isotope separation', Canadian Patent No. 1032666, 6 June 1978.

63. Tiee, J.J. and Wittig, C., 'A 0.1 J CF_4 laser operating in the 16 μm region', *Journal of the Optical Society of America,* Vol. 68, 1978, pp. 672–73.

64. Byer, R.L, 'A 16 μm source for laser isotope enrichment', *IEEE Journal of Quantum Electronics,* November 1976, pp. 732–33.

65. Rabinowitz, P. *et al.,* 'Stimulated rotational Raman scattering in H_2' (Exxon Research and Engineering Co., Linden, New Jersey) [undated preprint].

66. 'New infrared sources', *Laser Focus,* Vol. 17, No. 2, February 1981, pp. 76–78.

67. Brau, C.A., 'The free-electron laser: an introduction', *Laser Focus,* Vol. 17, No. 5, May 1981, pp. 48–56.

68. Ambartsumyan, R.V. *et al., ZhETF Pis'ma Red.,* Vol. 20, 1974, p. 587 [Engl. Trans. in *Soviet Physics JETP Letters,* Vol. 20, 1974, p. 273].

69. Tiee, J.J. and Wittig, C., 'The photodissociation of UF_6 using infrared lasers', *Optics Communications,* Vol. 27, No. 3, December 1978, pp. 377–80.

70. Cantrell, C.D., Louisell, W.H. and Lam, J.F., 'Coherent pulse propagation effects in multilevel molecular systems', in K.L. Kompa and S.D. Smith (eds.), *Laser-Induced Processes in Molecules* (Springer-Verlag, New York, 1979), pp. 138–41.

71. Bernard, P., Galarneau, P. and Chin, S.L., 'Self-focusing of CO_2 laser pulses in low-pressure SF_6', *Optics Letters,* Vol. 6, No. 3, March 1981, pp. 139–141.

72. Nowak, A.V. and Ham, D.O., 'Self-focusing of 10μm laser pulses in SF_6', *Optics Letters,* Vol. 6, No. 4, April 1981, pp. 185–87.

(a) —, p. 187.

73. Horsley, J.A. *et al.,* 'Laser chemistry experiments with UF_6', *IEEE Journal of Quantum Electronics,* Vol. QE-16, No. 4, April 1980, p. 418.

74. Rabinowitz, P., Stein, A. and Kaldor, A., 'Infrared multiphoton dissociation of UF_6', *Optics Communications,* Vol. 27, No. 3, December 1978, p. 384.

75. Emanuel, G., 'High enrichment steady-state cascade performance', *Nuclear Technology,* Vol. 43, No. 3, May 1979, pp. 314–27.

76. Emanuel, G., 'Cascade performance for large separation factors', *Nuclear Technology,* Vol. 51, No. 2, December 1980, pp. 238–43.

77. 'Isotope separation using vibrationally excited molecules', UK Patent Specification 1 537 757, 10 January 1979.

78. Winterberg, F., 'Combined laser centrifugal isotopic separation technique', *Atomkernenergie,* Vol. 30, 1977, pp. 65–66.

79. Lawande, S.V., and Lal, B., 'Selective momentum transfer—a novel technique in laser isotope separation', *Physics News,* June 1979, pp. 41–46.

80. Casper, B.M., 'Laser enrichment: a new path to proliferation?', *Bulletin of the Atomic Scientists,* Vol. 33, No. 1, January 1977, pp. 28–41.

81. Casper, B.M., 'Time for a moratorium' *Bulletin of the Atomic Scientists,* Vol. 33, No. 6, June 1977, pp. 54–56.

82. SIPRI, *World Armaments and Disarmament, SIPRI Yearbook 1982* (Taylor & Francis, London, 1982), chapter 12.

83. Hearings on H.R. 2969, Department of Energy Authorization Legislation (National Security Programs) for Fiscal Year 1982, before the Procurement and Military Nuclear Systems Subcommittee of the Committee on Armed Services, US House of Representatives, 97th Congress, 1st session, 2, 4, 5, 9 March 1981 (US Government Printing Office, Washington, D.C., 1981), pp. 141–76.

84. Spitzer, L., Jr., *Physics of Fully Ionized Gases* (Interscience Publishers Inc., New York, 1956), p. 17.

85. Lewis, H.W., 'Ball lightning', *Scientific American,* Vol. 208, No. 3, 1963, pp. 106–16.

86. Hewlett, R.G. and Anderson, O.E., Jr., *The New World 1939/1946* (Pennsylvania State University Press, University Park, Pa., 1962).
(a) —, pp. 91–96.
(b) —, pp. 142–43.
(c) —, p. 164.

87. Love, L.O., 'Electromagnetic separation of isotopes at Oak Ridge', *Science,* Vol. 182, No. 4110, 26 October 1973.
(a) —, pp. 343–52.
(b) —, p. 343.
(c) —, p. 344.
(d) —, p. 348.

88. Smith, M.L. and Hill, K.J., 'Operational Experience with Hermes; The Harwell Active Electromagnetic Separator', *Proceedings of the International Symposium on Isotope Separation,* Amsterdam, 1957 (North Holland, Amsterdam, 1958), pp. 581–95.

89. Dawson, J.M. *et al.,* 'Isotope separation in plasmas by use of ion cyclotron resonance', *Physical Review Letters,* Vol. 37, No. 23, 6 December 1976.
(a) —, pp. 1547–50.
(b) —, p. 1547.

90. Weibel, E.S., 'Separation of isotopes', *Physical Review Letters,* Vol. 44, No. 6, 11 February 1980, pp. 377–79.

91. Cohen, S.L., *Mission Analysis for Isotope Separation/Plasma Separation Process,* US Department of Energy Report, DoE/ET/33006–T1, June 1978, p. ii.

92. Vanstrum, P.R., *Statement before the Subcommittee on Energy Research and Production of the Committee on Science and Technology,* US House of Representatives (US Government Printing Office, Washington, D.C., 1979), 22

References to Part Two (cont.)

September 1979, p. 146.

93. Bonnevier, B., 'Experimental evidence of element and isotope separation in a rotating plasma', *Plasma Physics,* Vol. 13, 1971, p. 765.

94. Lehnert, B., 'A partially ionized plasma centrifuge', *Physica Scripta,* Vol. 2, 1970, pp. 106–107.

95. Bonnevier, B., 'Diffusion due to ion–ion collisions in a multicomponent plasma', *Arkiv für Fysik,* Vol. 33, No. 15, 1966.

(a) —, p. 267.

(b) —, p. 268.

96. James, B.W. and Simpson, S.W., 'Isotope separation in the plasma centrifuge' *Plasma Physics,* Vol. 18, 1976, pp. 289–300.

97. Belorusov, A.V. *et al.,* 'Separation of gas mixtures and xenon isotopes in a pulsed plasma centrifuge', *Soviet Journal of Plasma Physics,* Vol. 5, No. 6, November-December 1979, p. 696.

98. Anderson, J.B., 'Isotope separation in crossed jet systems', *AIChE Symposium Series,* Vol. 76, No. 192, 1980, pp. 89–92.

99. Wang, C.G., 'Garden hose separation of gaseous isotopes', *Nature,* Vol. 253, 24 January 1975, pp. 260–62.

100. Wang, C.G., 'Garden hose separation of gaseous isotopes, part II: supersonic accelerations', *Nuclear Technology,* Vol. 42, January 1979, pp. 90–101.

101. Gazda, H.O.E., 'The HAGA radial-separating-nozzle centrifuge process for uranium enrichment', *Atomkernenergie,* Vol. 31, 1978, pp. 4–5.

102. Gazda, H.O.E., 'Enrichment of uranium with the special emphasis on the HAGA radial-separating-nozzle centrifuge' *Atomkernenergie,* Vol. 33, 1979, pp. 2–4.

103. 'Enrichment at the speed of sound', *Nuclear Engineering International,* Vol. 26, No. 312, April 1981, pp. 44–45.

104. Lester, R.K., 'Secrecy, patents and non-proliferation', *Bulletin of the Atomic Scientists,* Vol. 37, No. 5, May 1981, pp. 35–38.

105. Hecht, J., 'America picks lasers to enrich uranium', *New Scientist,* Vol. 94, No. 1307, 27 May 1982, p. 554.

References to Part Three

1. Sherwin, M.J., *A World Destroyed: The Atomic Bomb and the Grand Alliance* (Vintage Books, New York, 1977), pp. 6 and 90–98.

2. Jungk, R., 'Brighter than a thousand suns' in *The Franck Report* (Victor Gollancz Ltd, London, 1958), pp. 335–46.

3. Epstein, W., *The Last Chance* (The Free Press, New York, 1976).
   (a) —, p. 5.
   (b) —, p. 8.
   (c) —, p. 16.
   (d) —, p. 237.

4. Herken, G., *The Winning Weapon: The Atomic Bomb in the Cold War 1945–50* (Knopf, New York, 1980).

5. Kramish, A., *Atomic Energy in the Soviet Union* (Stanford University Press, Stanford, Connecticut, 1959), pp. 110, 115.

6. Goldschmidt, B., *Le Complexe Atomique* (Fayard, Paris, 1980).
   (a) —, p. 103.
   (b) —, pp. 106–16.
   (c) —, pp. 156–57.
   (d) —, p. 153.
   (e) —, p. 156.
   (f) —, p. 313.
   (g) —, pp. 205, 315.
   (h) —, pp. 162–63.
   (i) —, p. 191.
   (j) —, p. 388.

7. Nieburg, H.L., *Nuclear Secrecy and Foreign Policy* (Public Affairs Press, Washington, D.C., 1964), chapter 5 and p. 68.

8. Scheinman, L., *Atomic Energy Policy in France under the Fourth Republic* (Princeton University Press, Princeton, New Jersey, 1965).
   (a) —, p. 86.
   (b) —, p. 173.
   (c) —, p. 176.
   (d) —, p. 196.
   (e) —, pp. 177–82.
   (f) —, chapters 4, 5.
   (g) —, pp. 137–38.
   (h) —, chapter 5, p. 185.

9. Kohl, W.L., *French Nuclear Diplomacy* (Princeton University Press, Princeton, New Jersey, 1971).
   (a) —, pp. 55–64.
   (b) —, pp. 54–61.
   (c) —, p. 55.

10. Office of Technology Assessment, *Nuclear Proliferation and Safeguards* (Praeger, New York, 1977).
    (a) —, p. 180.
    (b) —, p. 218.

11. Goldschmidt, B., 'International nuclear collaboration and Article IV of the Non-Proliferation Treaty', in SIPRI, *Nuclear Proliferation Problems* (MIT Press, Cambridge, Massachusetts, 1974).

References to Part Three (cont.)

(a) —, p. 211.

(b) —, p. 212.

12. Stikker, D.U., (former NATO Secretary General), in *de Nieuwe Rotterdamse Courant,* 19 April 1975, p. 7.

13. Nau, H.R., *National Politics and International Technology* (Johns Hopkins University Press, Baltimore, Maryland, 1974).

(a) —, p. 96.

(b) —, p. 155.

(c) —, pp. 155–56.

14. Euratom Treaty, Rome, 25 March 1957 (Verdrag ter oprichting van de Europese Gemeenschap voor Atoomenergie [Euratom], N.V. Uitgeversmij W.C.J. Tjeenk Willink, Zwolle, 1967), chapter VII.

15. INFCIRC/193, IAEA, Vienna, 18 April 1977.

16. 'Agreement between Belgium, Denmark, the Federal Republic of Germany, Ireland, Italy, Luxembourg, the Netherlands, the European Economic Community and the International Atomic Energy Agency in implementation of Article III, (1) and (4) of the Treaty on the non-proliferation of nuclear weapons 1973', Part I, in SIPRI, *Arms Control* (Taylor & Francis, London, 1978), pp. 106–109.

17. Schleicher, H.W., 'Nuclear safeguards in the European Community: a regional approach', *IAEA Bulletin,* Vol. 22, No. 3, 4 August 1980.

18. Mendershausen, H., *International Cooperation in Nuclear Fuel Services: European and American Approaches,* RAND/P-6308 (Rand Corp., Santa Monica, Cal., 1978).

(a) —, p. 31.

(b) —, p. 34.

(c) —, p. 11.

19. 'Agreement between the Kingdom of the Netherlands, the Federal Republic of Germany and the United Kingdom of Great Britain and Northern Ireland on collaboration in the development and exploitation of the gas centrifuge process for producing enriched Uranium', *Almelo Treaty,* Article II, *in Tractatenbladen van het Koninkrijk der Nederlanden,* jaargang 1970, no. 41 (Staatsuitgeverij, Den Haag, 1970).

20. Wonder, E.F., *Nuclear Fuel and American Foreign Policy: Multilateralization for Uranium Enrichment,* Atlantic Council Policy Paperback (Westview Press, Boulder, Colorado, 1977).

(a) —, pp. 11–25, 392.

(b) —, p. 43.

21. Kissinger, H., 'Building International Order', Address to the United Nations, 22 September 1975, *US Department of State Bulletin,* 13 October 1975.

22. SIPRI, *Internationalization to Prevent the Spread of Nuclear Weapons* (Taylor & Francis, London, 1980), p. 95.

23. SIPRI, *The NPT: The Main Political Barrier to Nuclear Weapon Proliferation* (Taylor & Francis, London, 1980), p. 51.

24. IAEA, *Regional Nuclear Fuel Cycle Centers,* Vol. I, (IAEA, Vienna, 1977).

(a) —, p. 3.

(b) —, chapters 4, 9.

25. SIPRI, *Nuclear Energy and Nuclear Weapon Proliferation* (Taylor &

References to Part Three (cont.)

Francis, London, 1979).
   (a) —, W.H. Donnelly, 'Application of US non-proliferation legislation for technical aspects of the control of fissionable materials in non-military applications', Paper No. 14, pp. 216–17.
   (b) —, W.H. Donnelly, p. 213.
   (c) —, W.H. Donnelly, pp. 199–222.
   26. Moore, T.G., *Uranium Enrichment and Public Policy,* AEI-Hoover Policy Studies 25 (American Enterprise Institute for Public Policy Research, 1978), p. 36.
   27. MITRE Corporation, *Nuclear Power: Issues and Choices* (Ballinger, Cambridge, Massachusetts, 1977).
   (a) —, pp. 36, 374–75.
   (b) —, p. 299.
   28. SIPRI, *World Armaments and Disarmament, SIPRI Yearbook 1976* (Almqvist & Wiksell, Stockholm, 1976), p. 389.
   29. SIPRI, *World Armaments and Disarmament, SIPRI Yearbook 1977* (Almqvist & Wiksell, Stockholm, 1977).
   (a) —, p. 33.
   (b) —, p. 362.
   (c) —, p. 35.
   (d) —, p. 20.
   30. SIPRI, *World Armaments and Disarmament, SIPRI Yearbook 1978* (Taylor & Francis, London, 1978).
   (a) —, p. 35.
   (b) —, pp. 69–79.
   (c) —, pp. 70–73.
   31. Nuclear Suppliers Group, *Guidelines for Nuclear Transfer,* 1978. Guidelines including annexes A & B (available from the Netherlands Department of Foreign Affairs, The Hague).
   32. Casper, B.M., *Bulletin of the Atomic Scientists,* Vol. 33, No. 1, January 1977, pp. 28–41.
   33. Presidential Documents Jimmy Carter 1977; Nuclear Power Policy (9 April 1977) in *Legislative History of the Nuclear Non-Proliferation Act of 1978,* H.R. 8638 (Public Law 95–242), prepared for the Subcommittee on Energy, Nuclear Proliferation and Federal Services, Committee on Governmental Affairs, US Senate, by the Congressional Research Service (Library of Congress, Washington, D.C., January 1978), p. 933.
   34. Nuclear Non-Proliferation Act of 1978 (Public Law 95–242), 10 March 1978.
   (a) —, Sections 304, 306, 401.
   (b) —, Section 101.
   (c) —, Section 104.
   35. *International Nuclear Fuel Cycle Evaluation (INFCE)* (IAEA, Vienna, 1980).
   (a) —, INFCE/PC/2/4, 'Reprocessing, plutonium handling, recycle', p. 150.
   (b) —, INFCE/PC/2/2, 'Enrichment availability', p. 127.
   (c) —, *ibid.,* p. 147.
   (d) —, *ibid.,* p. 24.
   (e) —, *ibid.,* p. 146.
   (f) —, INFCE/PC/2/9, 'Summary volume', p. 51.

References to Part Three (cont.)

(g) —, *ibid.*, p. 38.

(h) —, *ibid.*, p. 44.

(i) —, INFCE/PC/2/2, 'Enrichment availability', p. 74.

(j) —, *ibid.*, p. 52.

36. *Proceedings of the International Conference on Uranium Isotope Separation*, London, 5–7 March 1975 (unpublished).

(a) —, Hill, J.H., 'Uranium enrichment in the United States'. Paper No. 27.

(b) —, Jetter, H., Gürs, K., Dibbert, H.J., 'Uranium isotope separation using IR lasers', Paper No. 4.

(c) —, van der Spek, J.H., 'Project for a uranium enrichment plant at Inga (Zaire)', abstract.

37. *Energy and Water Development Appropriations, 1980*, Part 6, US Senate Hearings, No. 45–638 (US Government Printing Office, Washington, D.C., 1979).

(a) —, p. 1409.

(b) —, p. 1524.

(c) —, p. 1444.

38. 'Centrifuges key to U-enrichment future', *Nuclear News,* Vol. 20, No. 8, June 1977, p. 66.

39. *International Conference on the Nuclear Fuel Cycle* (Atomic Industrial Forum/Dutch Atomic Forum, Amsterdam, 14–17 September 1980, unpublished).

(a) —, Davis, R.M., 'Commercial enrichment services — supply assurance; US Department of Energy's views as a supplier', pp. 5–6.

(b) —, Jelinek-Fink, P., 'URENCO's view of the enrichment market', p. 15.

(c) —, Oshima, K., 'The status and future prospects of uranium enrichment in Japan', pp. 6–9.

(d) —, Oshima, K., pp. 3–4.

(e) —, Oshima, K., p. 5.

(f) —, Oshima, K., p. 8.

(g) —, Oshima, K., p. 9.

(h) —, Barrat, P.H., 'Assurance of uranium supply: Australian development and export policy and perspectives', p. 24.

40. 'World list of nuclear power plants', *Nuclear News,* Vol. 24, No. 2, February 1981, pp. 75–94.

41. 'Uranium (resources, production and demand)', *Joint Report OECD/IAEA* (OECD, Paris, 1979), pp. 18, 22, 23.

42. Voight, W.R., Jr., 'Uranium enrichment: response to an energy crisis', *Fuel Cycle Conference '80* (Atomic Industrial Forum, New Orleans, Louisiana, 15–18 April 1980, unpublished).

(a) —, p. 6.

(b) —, p. 7.

43. 'Advanced uranium enrichment technologies', *Hearing before the Subcommittee on Energy Research and Production,* Committee on Science and Technology, US House of Representatives, No. 58 (US Government Printing Office Document No. 54–835, Washington, D.C., 1979).

44. *West European Nuclear Energy Development: Implications for the United States,* Congressional Research Service, USGPO No. 46–183 (US Government Printing Office, Washington, D.C., 1979).

(a) —, p. 8.

(b) —, pp. 72–73.

References to Part Three (cont.)

(c) —, p. 100.

(d) —, pp. 79–80.

(e) —, p. 78.

(f) —, p. 82.

(g) —, p. 83.

(h) —, p. 62.

(i) —, pp. 88–91.

(j) —, p. 87.

45. Duffy, G., *Soviet Nuclear Energy: Domestic and International Policies,* R-2362–DoE (US Department of Energy, Washington, D.C., December 1979).

(a) —, pp. 67–72.

(b) —, pp. 5–13.

(c) —, p. 7.

(d) —, p. 17.

(e) —, pp. 23–24.

46. *Bulletin of the Atomic Scientists,* Vol. 37, No. 2, February 1981.

(a) —, F. Barnaby, 'A nuclear engineer's paradise', pp. 55–56.

(b) —, H. Krugman, 'The German–Brazilian nuclear deal', pp. 32–36.

47. *Second Review Conference of the Parties to the Treaty on the Non-Proliferation of Nuclear Weapons,* Geneva, 12 August 1980, (unpublished).

(a) —, I.G. Morozov, 'Statement by the head of the USSR delegation', p. 9.

(b) —, D.S. McPhail, 'Statement by the head of the Canadian delegation', p. 9.

48. *World Nuclear Power and its Fuel Cycle after Three Mile Island: A Guide for Marketing and Strategic Planning* (Interdevelopment, Inc., Arlington, Virginia, April 1980).

(a) —, p. III-126.

(b) —, pp. III-131, 132.

(c) —, p. III-135.

(d) —, p. III-124.

49. 'Italien reduziert Eurodif-Anteil', *Atomwirtschaft/Atomtechnik,* Vol. 25, No. 6, June 1980.

50. 'Eurodif-Anlage erreicht 6 Mio. UTA/a', *Atomwirtschaft/Atomtechnik,* Vol. 25, No. 10, October 1980.

51. 'Progressive start up of the Tricastin . . . ', *Nuclear News,* Vol. 22, No. 5, p. 23/113.

52. Lurf, G., 'Die Französischen Aktivitäten im Brennstoff-kreislauf', *Atomwirtschaft/Atomtechnik,* Vol. 24, No. 2, February 1979, pp. 76–82.

53. *Nuclear Power and its Fuel Cycle,* Proceedings of an International Conference, Salzburg, 2–13 May 1977, Vol. 3 (IAEA, Vienna, 1977).

(a) —, C. Fréjacques *et al.,* 'Evolution des procédés de separation des isotopes de l'uranium en France', pp. 203–13.

(b) —, P. Bullio *et al.,* 'Italian activities in uranium enrichment', pp. 183–201.

54. Blumkin, S., *Survey of Foreign Enrichment Capacity, Contracting and Technology,* K/OA-2547 (Part 5), 30 October 1978.

(a) —, p. 16.

(b) —, p. 18.

(c) —, p. 15.

55. SIPRI, *World Armaments and Disarmament, SIPRI Yearbook 1972* (Almqvist & Wiksell, Stockholm, 1972).

    (a) —, p. 334.
    (b) —, p. 328.
    (c) —, pp. 489–500.
    (d) —, p. 316.
    (e) —, p. 308.
    (f) —, p. 310.

56. Lefever, E.W., *Nuclear Arms in the Third World* (The Brookings Institution, Washington, D.C., 1979).
    (a) —, p. 46.
    (b) —, pp. 106–107.
    (c) —, p. 42.
    (d) —, p. 45.
    (e) —, p. 43.
    (f) —, pp. 27–28.
    (g) —, p. 26.
    (h) —, p. 31.
    (i) —, p. 36.
    (j) —, p. 37.
    (k) —, p. 69.
    (m) —, p. 86.
    (n) —, p. 89.
    (o) —, p. 88.

57. *Nuclear Proliferation Factbook,* Congressional Research Service (US Government Printing Office, Washington, D.C., September 1980) No. 95–041.
    (a) —, pp. 190–92.
    (b) —, p. 173.
    (c) —, p. 219.
    (d) —, p. 193.

58. *Energy and Water Development Appropriations, Fiscal Year 1981,* Part 2, US Senate Hearings, US GPO No. 65–651, (US Government Printing Office, Washington, D.C., 1980), p. 1088.

59. 'Italy wants to trim its Eurodif share . . . ', *Nuclear News,* Vol. 23, No. 7, May 1980, pp. 20–89.

60. SIPRI, *World Armaments and Disarmament, SIPRI Yearbook 1979* (Taylor & Francis, London, 1979), pp. 314–19.

61. 'Iraq intends to use nuclear power only for peaceful purposes', *Nucleonics Week,* Vol. 21, No. 30, 24 July 1980, pp. 4–5.

62. 'Egyptians and Libyans ready to allow international scrutiny of atomic plants', *International Herald Tribune,* 2 March 1981, p. 4.

63. Mathieu, P., 'Laser uranium isotope separation techniques', University of Liège, Faculty of Applied Science, College Publication (1977), No. 65, pp. 135–62, abstract *in INIS Atomindex* no. 445444, Vol. 10, No. 8, 15 April 1979 (IAEA, Vienna), p. 2959.

64. Mathieu, P., 'Mathematical model of non-equilibrium condensation in one- and two-dimensional two-phase flows', *Letters in Heat and Mass Transfer,* Vol. 6, No. 1, January 1979, pp. 61–71.

65. Mathieu, P., 'Laser isotope separation by selective chemical reactions in two-phase flows', *Letters in Heat and Mass Transfer,* Vol. 6, No. 4, July 1979, pp. 329–34.

66. 'Belgium to decide fate of Eurochemic reprocessing plant early in 1981', *Nucleonics Week,* Vol. 21, No. 46, 13 November 1980, p. 11.

67. *Nucleonics Week,* Vol. 21, No. 36, 4 September 1980.

(a) —, 'Iran must reinstate its scrapped nuclear power program', p. 10.

(b) —, 'Canadians expect nuclear business deals to follow visit by Chinese', p. 7.

68. 'Iran: fate of nuclear projects in doubt in new regime', *Nuclear News,* Vol. 22, No. 3, March 1979, p. 59.

69. 'Iran: Kraftwerk Union withdraws', *Nuclear News,* Vol. 22, No. 11, September 1979, p. 52.

70. 'Iranian attack on Iraqi nuclear complex seen as harbinger', *Nucleonics Week,* Vol. 21, No. 41, 9 October 1980, p. 1.

71. 'Iran drops role in nuclear fuel plant', *International Herald Tribune,* 18 February 1980.

72. 'Téhéran veut rester dans Eurodif', *Le Monde,* 26 February 1980, p. 43.

73. 'La société Technicatome retient en France 80 tonnes d'uranium Iranien', *Le Monde,* 23 April 1981, p. 40.

74. *Jahrbuch der Atomwirtschaft 1981* (Handelsblatt GmbH, Düsseldorf, FRG, 1981).

(a) —, 'Luftangriff auf Kernforschungszentrum', p. 159.

(b) —, 'Unklare Kernkraftswerkpläne', p. 126.

75. 'Uranit plant Anreicherungsanlage', *Atomwirtschaft/Atomtechnik,* Vol. 21, No. 12, December 1976, pp. 557–58.

76. *Atomwirtschaft/Atomtechnik,* Vol. 20, No. 7/8, July/August 1975.

(a) —, 'Das Deutsch-Brazilianische Abkommen', pp. 321–22.

(b) —, 'Anreicherungsplane im Zweifel', p. 324.

77. Červenka, Z. and Rodgers, B., *The Nuclear Axis* (Julian Freedmann Books, London, 1978).

(a) —, pp. 116–55.

(b) —, p. 136.

78. Smith, D., *South Africa's Nuclear Capability* (World Campaign Against Military and Nuclear Collaboration with South Africa and UN Centre Against Apartheid, London, February, 1980).

79. *Atomwirtschaft/Atomtechnik,* Vol. 26, No. 2, February 1981.

(a) —, P. Bauder, 'Entwicklung des Uranbergbaus in Australien', pp. 84–88.

(b) —, 'Französisch-Brazilianischer UF_6-Vertrag in Kraft', p. 57.

80. Geoghegan, G.R.H. and Kehoe, R.B., 'Uranium enrichment in the UK', *Atom,* Vol. 169, November 1970, pp. 224–30.

81. 'Nota inzake het Kernenergiebeleid', *Kamerstuk* (Report to Dutch Parliament), 11761, No. 1, 30 March 1972, p. 2.

82. 'China is planning a big nuclear power program', *Nucleonics Week,* Vol. 21, No. 47, 20 November 1980, p. 7.

83. 'China, Hong Kong move toward decision on acquiring nuclear unit', *Nucleonics Week,* Vol. 21, No. 25, 25 June 1980, pp. 9–10.

84. 'France will recommend discussions to supply China its first nuclear station', *Nucleonics Week,* Vol. 21, No. 43, 23 October 1980., p. 10.

85. 'China and Hong Kong aiming for two-unit nuclear station', *Nucleonics Week,* Vol 21, No. 39, 25 September 1980, p. 1.

86. 'Mainland Chinese talking to French, Germans, about nuclear power',

References to Part Three (cont.)

Nucleonics Week, Vol. 19, No. 2, 12 January 1978, p. 3.

87. Murphy, C.H., 'Mainland China's evolving nuclear deterrent', *Bulletin of the Atomic Scientists,* January 1972, p. 28.

88. 'China, Kernbrennstoff-Aktivitäten', *Atomwirtschaft/Atomtechnik,* Vol. 25, No. 12, December 1980, p. 596.

89. 'Potential customers from China visit Nuclex', *Nuclear News,* Vol. 21, No. 14, November 1978, p. 72.

90. 'Nuclear weapons history: Japan's wartime bomb project revealed', *Science,* Vol. 199, 1978, pp. 152–57.

91. 'Erfolgreiche Urananreicherung mit Laserstrahlen', *Laser+Elektro-Optik,* No. 4/1980, p. 20.

92. 'Centrifuge plant to start this month', *Nuclear News,* Vol. 22, No. 2, February 1979, p. 90.

93. 'UCOR-Versuchsanlage in Betrieb', *Atomwirtschaft/Atomtechnik,* Vol. 20, No. 5, May 1975, p. 226.

94. 'The South African Government has announced . . .', *Nuclear News,* Vol. 21, No. 4, March 1978, p. 20.

95. *Nuclear Fuel Cycle Requirements (and Supply Considerations through the Long Term)* (OECD, Paris, February 1978).

(a) —, p. 32.

(b) —, p. 74.

96. Becker, E.W., Nogueira Batista, P. and Völcker, H., 'Uranium enrichment by the separation nozzle method within the framework of German / Brazilian cooperation', *Nuclear Technology,* Vol. 52, No. 1, January 1981, pp. 105–14.

97. SIPRI, *World Armaments and Disarmament, SIPRI Yearbook 1980* (Taylor & Francis, London, 1980), Chapter 17, 'The implementation of multilateral arms control agreements'.

(a) —, pp. 443–68.

(b) —, p. 497.

98. 'Brazil and Argentina signed several nuclear cooperation accords', *Nucleonics Week,* Vol. 21, No. 21, 22 May 1980, p. 12.

99. Series of articles about "The Islamic Bomb", *de Volkskrant* in cooperation with *The New York Times,* 21, 23 & 26 June 1980.

100. 'Atombomben für den Islam?', *Der Spiegel,* 12 November 1979, pp. 202–209.

101. 'How Dr. Khan stole the bomb for Islam', *The Observer,* 9 December 1979.

102. Sinha, P.B. and Subramanian, R.R., *Nuclear Pakistan* (Vision Books Private Ltd., New Delhi, 1980), p. 121.

103. Lodgaard, S., 'Nuclear proliferation: critical issues', *Bulletin of Peace Proposals,* Vol. 1, 1981, pp. 11–20.

104. SIPRI, *World Armaments and Disarmament, SIPRI Yearbook 1975* (Almqvist & Wiksell, Stockholm, 1975), pp. 438–44.

105. *Asian Survey,* Vol. 20, No. 5, May 1980.

(a) —, A. Kapur, 'A nuclearizing Pakistan: some hypotheses', pp. 495–516.

(b) —, R. Tomar, 'The Indian nuclear power program: myths and mirages', pp. 517–31.

106. Kapur, A., 'The Canada–India nuclear negotiations: some hypotheses

References to Part Three (cont.)

and lessons', *The World Today*, August 1978, pp. 311–20.

107. Power, P.F., 'The Indo-American nuclear controversy', *Asian Survey*, Vol. 19, No. 6, June 1979, pp. 582–83.

108. 'Indian enrichment experiments ostensibly for weapons purposes continue', *Nucleonics Week*, Vol. 21, No. 15, 10 April 1980, p. 12.

109. 'Ranger-Abbau kann beginnen', *Atomwirtschaft/Atomtechnik*, Vol. 22, No. 9, September 1977, p. 446.

110. Salaff, S., 'Bar sinister. The Anglo-Dutch-West German consortium for the enrichment of uranium', *Current Research on Peace and Violence*, Vol. 1, No. 3–4 (Tampere Peace Research Institute, Tampere, Finland), 1978, p. 157.

111. Davenport, E., Eddy, P. and Gillman, P., *The Plumbat Affair* (J.B. Lippincott Company, New York, 1978).

112. 'Dayan says Israel can make nuclear weapons quickly', *International Herald Tribune*, 25 June 1981, p. 1.

113. 'Special report, how Israel got the bomb', *Time*, 12 April 1976, p. 39.

114. 'Richtlinien für Uranexport', *Atomwirtschaft/Atomtechnik*, Vol. 22, No. 1, January 1977, p. 8.

115. Boskma, P., Smit, W. and de Vries, G., *Uranium Verrijking*, Boerderij-cahier 7501 (Twente University of Technology, Enschede, the Netherlands), June 1975.

116. 'Eurodif ohne Schweden', *Jahrbuch der Atomwirtschaft 1975* (Handelsblatt GmbH, Düsseldorf, FRG, 1975).

117. 'Les Sud-Coréens estiment que les Français sont plus libéraux que les Américains en matière de transferts de technologie', *Le Monde*, 8 April 1981.

118. Young-sun, H., 'Nuclearization of small states and world order: the case of Korea', *Asian Review*, Vol. 18, 1978, p. 1142.

119. 'A new antinuclear movement is being directed in Switzerland . . . ', *Nucleonics Week*, Vol. 21, No. 33, 14 August 1980, p. 4.

120. 'Niger verkoopt desnoods uranium aan de duivel', *Volkskrant*, 17 April 1981.

121. *Uranium Industry Seminar*, Department of Energy, Grand Junction, Co., USA, *GJO-108*, 17 and 18 October 1978, pp. 68–71.

122. 'Algerien: Erschliessung von Uranvorkommen', *Atomwirtschaft/Atomtechnik*, Vol. 25, Nos. 8–9, August/September 1980, pp. 402–403.

123. 'Algeria is planning to launch a nuclear program', *Nucleonics Week*, Vol. 22, No. 10, 12 March 1981, p. 6.

124. Wilcox, W.J., Jr., 'Uranium enrichment — a review of the present world status: capacity, technology and plans', K/TD-394, 15 February 1979, p. 9.

125. 'Mitterrand zur Kernenergie', *Atomwirtschaft/Atomtechnik*, Vol. 26, No. 6, June 1981, p. 333.

126. 'La politique nucléaire', *Le Monde*, August 1, 1981, p. 1.

127. 'Bau des Brüters in Kalkar höchst gefährdet', *Frankfurter Algemeine Zeitung*, 24 August 1981.

128. Mohrhauer, H., Krey, M. and Severin, D., 'Urananreicherung mit Zentrifugen; die deutsche Anlage und das Ausbauprogramm der Urenco', *Atomwirtschaft/Atomtechnik*, Vol. 26, No. 3, March 1981.

(a) —, p. 189.

(b) —, p. 186.

129. 'Pakistan is said to build secret A-plant', *International Herald Tribune*,

24 September 1980, p. 4.

130. Norman, C., 'Gas centrifuge plant in trouble in Congress', *Science,* Vol. 216, 11 June 1982, pp. 1206–1207.

131. 'A bomb pledge by Gandhi', *DAITEL,* 11 July 1981.

132. Rajya Sabha, Department of Atomic Energy, Government of India, Starred Question No. 49 to Prime Minister Indira Gandhi, 19 February 1981.

133. 'India not making N-bomb', *National Herald,* 31 December 1980.

134. *Nucleonics Week,* 16 July 1981, p. 8.

135. '14 Kerncentrales extra nodig voor EG', *NRC-Handelsblad,* 15 April 1980.

136. 'Publieke opinie over kernenergie in Japan', *Tokio Nieuws,* No. 132, Dutch Department of Economic Affairs, April 1981.

137. 'World list of nuclear power plants', *Nuclear News,* Vol. 22, No. 10, August 1979, pp. 69–87.

138. *Fuel and Heavy Water Availability,* INFCE, Report of Working Group 1 (IAEA, Vienna, 1980), pp. 44–47.

139. 'Betriebserfahrungen mit Kernkraftwerken', *Jahrbuch der Atomwirtschaft 1979* (Handelsblatt Verlag GmbH, Düsseldorf, FRG, 1979), p. B 25.

140. Pelser, J., 'De verrijking van uranium', *Atoomenergie,* June 1975, p. 138.

141. *International Conference on Uranium Enrichment, New Orleans, Louisiana, 29 Jan–1 Feb 1978* (Atomic Industrial Forum).

(a) —, J. Asyee, 'Urenco enrichment services', p. 20.

(b) —, J. F. Petit, 'Uranium enrichment by Eurodif/Coredif', p. 7.

142. Civiak, R.L., *Uranium Enrichment Services: Supply and Demand* (US Congressional Research Service, Library of Congress, 4 March 1981), p. CRS-6.

143. *Enrichment Availability, INFCE, Report of Working Group 2* (IAEA, Vienna, 1980), pp. 74–75.

144. *Department of Energy Authorizations for Fiscal Year 1981,* Congress Hearings, Serial No. 96–33, US GPO No. 66–885 (US Government Printing Office, Washington, D.C., 1980), p. 102.

145. *Uranium: Resources, Production and Demand,* Joint Report OECD/ IAEA (OECD. Paris, 1979), p. 26.

146. 'US rules led Australia to reject atomic deal', *International Herald Tribune,* 8 October 1982, p. 4.

Index